Lars Håkanson

A Manual of
Lake Morphometry

With 49 Figures

Springer-Verlag
Berlin Heidelberg New York 1981

Dr.Lars Håkanson
National Swedish Environment Protection Board
Water Quality Laboratory
Box 8043, S-75008 Uppsala, Sweden

ISBN 3-540-10480-1 Springer-Verlag Berlin Heidelberg New York
ISBN 0-387-10480-1 Springer-Verlag New York Heidelberg Berlin

Library of Congress Cataloging in Publication Data.
Håkanson, Lars. A manual of lake morphometry.
Bibliography: p. Includes index. 1. Lakes. I. Title.
GB591.H3   551.48'2   80-27975.

Offsetprinting and bookbinding: Fotokop Wilhelm Weihert, Darmstadt.
2132/3130-543210

CONTENTS

A	the total area, i.e. the lake area plus the area of islands, islets and rocks (km^2);
A_i	the total area of islands, islets and rocks within the shoreline (km^2);
a	the lake area (km^2);
a_i	the total area (= the cumulative area) within the limits of the contour line l_i (km^2);
a'	$10^5 \cdot \log A$;
a''	the area between two given contour lines (km^2);
$\bar{B}$	the mean width (km);
B_e	the maximum effective width (km);
B_{max}	the maximum width (km);
C	concave lake form;
CTP	checkered transparent paper, method for determining, e.g. the shoreline length;
Cx	convex lake form;
$\bar{D}$	the mean depth (m);
D_{max}	the maximum depth (m);
D_r	the relative depth (%);
D_{25}	the 1:st quartile depth (m);
D_{50}	the 2:nd quartile depth or the median depth (m);
D_{75}	the 3:rd quartile depth (m);
E	the area error (km^2);
e	the base for natural logarithms (= 2.718);
F	the shore development;
$f(\bar{x})$	the mean lake form;
$f(0.5)$, $f(2.0)$ etc	the statistical deviation forms from the mean lake form;
I	the information value of the bathymetric map;
I_n	insulosity (%);
I'	correctly identified area in the bathymetric map;
I''	the information number of the bathymetric map, which depends on the number of contour lines (n);
k	the normalized correction factor;
kHz	kilohertz; the frequency of the echosounder;
K_1	$\log(s + a')$ for $s = 1$, the reference scale;

K_2	$\log(s + a')$ for $s = 6\ 000\ 000$, the scale constant;
L	linear lake form;
L^+	the total estimated length in km of all contour lines in the bathymetric map;
L_e	the maximum effective length (km);
L_f	the effective fetch (km);
L_{max}	the maximum length (km);
L_r	a measure of the intensity of the hydrographic survey (km);
L_r'	the uncorrected L_r-value;
L_s	the effective length (km);
l_o	the shoreline length (km);
l_o'	the scale dependent l_o-value;
l	the total length of the contour lines in the bathymetric map (km);
l_c	the contour-line interval (m);
l_i	the length of a given contour line in the bathymetric map (km);
l_i^+	the estimated length in km of the given contour line (l_i);
l_x	the real length along the echoprofile in the x-direction;
l_y	the real length along the echoprofile in the y-direction;
ma	(macro), label for lakes with no point of inflexion in the relative hypsographic curve;
me	(meso), label for lakes with one point of inflexion in the relative hypsographic curve;
mi	(micro), label for lakes with two or more points of inflexion in the relative hypsographic curve;
n	the number of contour lines in the bathymetric map;
n'	the number of necessary determinations with CTP-technique to obtain, e.g. the shoreline length with a given statistical certainty;
R	the lake bottom roughness;
R'	the lake bottom roughness;
S	the map scale, e.g. 1:50 000, 1:100 000;
S'	the scale factor, e.g. $S' = 2.5$ for $S = 1:250\ 000$ and $S' = 5.0$ for $S = 1:500\ 000$;
SCx	slightly convex lake form;
s	the scale factor, $s = 1/S$;
s_i	the straight-line length of a given echosounded track (km);
s_n	the total length of all echosounded tracks (s_i) in a lake (km);

s_x the total straight-line length of all echosounded tracks in the x-direction (km);

s_y the total straight-line length of all echosounded tracks in the y-direction (km);

V the lake volume (km^3);

VCx very convex lake form;

V_d the volume development;

V_1 the linear approximation of the lake volume (V);

V_p the parabolic approximation of the lake volume (V);

W_{0-1} the water content of surficial sediment (0-1 cm), given in percentage of the wet substance;

α the slope;

α^o the slope in degrees;

α_p the slope in %;

$\bar{\alpha}_p$ the mean slope of a lake (%);

α_{50} the median slope of a lake;

β_i the acute angle between echosounded tracks;

γ_i a given deviation angle from the central radial; formula (8) for calculations of the effective fetch;

π 3.1416.

1. INTRODUCTION

In this context morphology means the study of lake forms and form elements, their genesis (from geographical and geological viewpoints) and their role in a broad physical limnological perspective. Lake morphometry deals with the quantification and measurement of these forms and form elements.

Morphometric data are of fundamental importance in most limnological and hydrological projects. This is obvious to most scientists in the field, but it is just as evident that lake morphology and lake morphometry are comparatively neglected topics of scientific endeavour. One can take the classical work of Welch - "Limnological methods" - as an example. This book was published in 1948 and it is still used as the main reference (see e.g. Wetzel, 1975), in spite of the fact that the source of most morphometric data, the bathymetric map, today is practically always constructed from hydrographic surveys conducted with echosounding equipment, a technique that became widely used and accepted only after 1948. Echosounding is not even mentioned in Welch's book. One can also point out that all scale-dependent morphometric parameters, like the shoreline length, the shore development and the lake bottom roughness, up till quite recently, have had limited quantitative relevance, since these parameters could not be defined unambiguously. For example, the shoreline length of Lake Vänern is about 1 000 km if determined on a map in scale 1:1 000 000, about 1 600 km when measured on a map in scale 1:100 000, and about 1 900 km from a map in scale 1:50 000 (Håkanson, 1978a). What map scale, what length value should be utilized? It is first with the use of the normalization formula (equation 15) that these scale dependent parameters get a meaning per se. These are just two examples indicating the progress of science, since Welch published his excellent textbook.

Another motive for writing this manual has been that there is at present a remarkable confusion about the meaning and definition of many morphometrical parameters. For example, in the three standard textbooks, Welch (1948), Hutchinson (1957) and Wetzel (1975), we find three different names for one and the same thing - maximum effective length (Welch) - length (Hutchinson) - maximum length (Wetzel), i.e. the straight line connecting the two most distant points on the shoreline over which wind and waves may act without interruptions from land or islands. If this manual can help to create a better terminology for lake morphometry, then I think that something will have been gained. The main purposes for this work are:

1

- to give a short but practical description on how to conduct echosoundings in lakes;
- to describe how the echograms are evaluated and utilized to construct a bathymetric map for a lake;
- to give a brief manual how to optimize lake hydrographic surveys and determine the information value of a bathymetric map;
- to present a thorough discussion concerning definitions and determinations of morphometrical parameters;
- to give short examples of how the morphometrical parameters may be used in various practical limnological, hydrological and sedimentological contexts.

The intention is that this manual should cover the basics of lake morphometry, which ought to be discussed at university courses in limnology, hydrology, geology and physical geography. It is intended that this book will also be usable for technical personnel from institutes and governmental offices working with lake investigations.

In order to keep the manual a reasonable size I have focused on the bathymetric map for whole lakes. Those particular problems that may arise when working in bays, sub-basins or in coastal areas will not be particularly dealt with, since this would lead to many exceptional cases and a thick and detailed manual, which would not correspond to the normal basic course at the university level.

To obtain a comprehensive picture of how the various morphometrical parameters can be used, I have used one lake as a type lake, namely Lake Vänern, Sweden. All parameters will be discussed in general terms, but exemplified with data from Lake Vänern. Background information concerning certain relevant data for Lake Vänern and its drainage area is given in the Appendix.

2. ECHOSOUNDINGS

2.1. Introduction

This section is meant to give a general working guide on how to plan and per-
form hydrographic surveys for scientific purposes with echosounding equipment.
The ultimate goal of the survey should be a bathymetric map with a specified
level of information.

At the outset it should be emphasized that the concept hydrographic survey is a cluster name for many different things. Hydrographic surveys may be
conducted with different techniques and for various purposes, e.g. to meet
requirements for safe navigation. In such cases the accuracy of every detail
and depth value is essential. For fishing purposes it is of interest to know
the bottom roughness, if there are obstacles on the bottom that may damage
the fishing equipment, as well as positions of "fishing banks". For scientific purposes, in hydrological, limnological and sedimentological contexts,
it is of primary interest to have a bathymetric map illustrating the general
morphology of the basin as well as major topographical features like trench-
es, sub-aquatic tops and deep-holes, topographical "bottle-necks" etc.

2.2. Instrumentation

Sounding equipment is manufactured commercially both for fixed installations,
generally on comparatively large boats and ships, and for mobile use. The
mobile systems are generally suitable for small motor-boats for use in lakes,
rivers and coastal areas. One type of mobile system is illustrated on Fig. 1.

The functional principle of echosounding is illustrated schematically on
Fig. 2. The echosounding equipment consists basically of: a transmitter, a
sender- and receiver unit for the acoustic wave, an amplifier unit, a recording unit and a mechanical suspension attachment. Most instruments are fed by
12 volts DC-power from an accumulator, and work with frequencies in the range
15 - 250 kHz; low frequencies (<50 kHz) provide better sediment penetration
and hence also information on physical sediment character than do high frequencies.

The transmitter produces pulses of a certain given frequency. The crystal
in the under-water sender-receiver unit is fed by these pulses and forced
into vibration. The vibrations are propagated in the water as acoustic waves.
The acoustic waves are sent out in the form of a cone, and not a straight

3

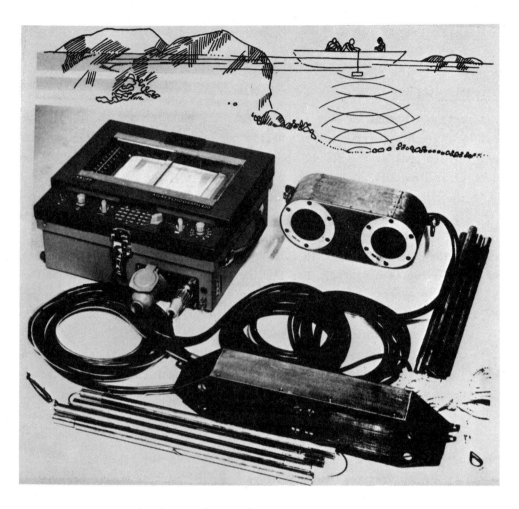

Fig. 1. One type of mobile echosounding equipment

line. The base of the cone gets wider with water depth (Fig. 3). This implies that the reflected signal gets weaker with water depth. The reflected signal is caught by the receiver, amplified and registered on the echogram as a depth value. The acoustic wave has a velocity of about 1.5 km/sec. in salt water. The echosounding equipment measures in reality the time it takes for the sonic wave to go down and up, but since the velocity is practically constant, the recording unit is graded in length units and not time units. Some instruments can be calibrated for water temperature and salinity by means of a frequency meter affecting the velocity of the recording pen.

The sender-receiver packet should always be placed in the middle of the boat and at about 1.5 m depth, to prevent it from coming above the water

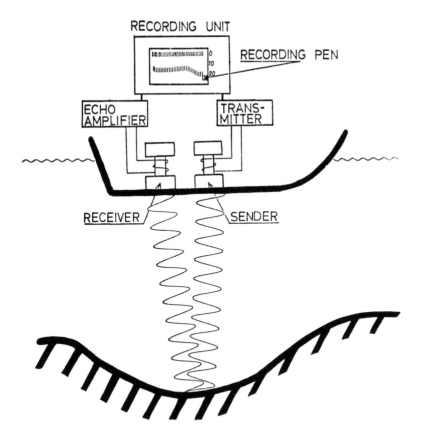

BOTTOM REFLECTION

Fig. 2. Outline of the functional principle of an echosounding equipment

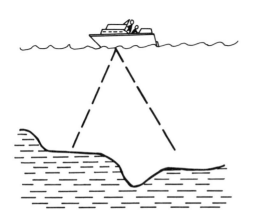

Fig. 3. Schematic illustration of the conic form of the ultra-sonic pulse

surface during the survey. The zeroline on the echogram should be adjusted to this depth value. The under-water unit should not be placed beside a keel or a fen where the acoustic wave may be shadowed, or at places where air-bubbles may affect the quality of the signal. It is of utmost importance to see that the sender-receiver packet is unaffected by vegetation etc, that may reduce the functional ability.

Complementary field equipment on a hydrographic survey is: a graduated sounding line with a lead weight, a pair of field glasses, a tool set, drawing materials etc. In shallow water areas, with depth less than 3 m, it is often advisable to use a sounding rod. The motor-boat should preferably be equipped with a good navigation compass, for proper steering, and a log to measure the distance.

2.3. Preparations

Working map; this should be an accurate map, generally in the best available scale, over the survey area. For practical purposes it is advantageous if the working map is coated with a transparent plastic film, making it water-proof. If several working maps are needed and used for one survey area, it is advisable to make a general working map in smaller scale for the whole area. If one has access to areal photos over the survey area, important information about conditions in the shallow water areas, stones, small piers, houses, boundary clearing etc, may be gained.

Sounding tracks; i.e. the lines along which the echosoundings are executed. These course lines should be drawn in advance on the working map. If possible, they should be placed in such a way that the terminal points can be sought out from one end to the other. The track net should preferably provide an even area cover of the survey area. It is advantageous if the tracks are parallel and placed perpendicular to the shoreline.

To optimize the yield of the survey in terms of accuracy it is recommended that a system of tracks crossing the main tracks system is used - preferably at right angles. The intensity of the survey, i.e. the average distance between the tracks (the L_r-value, see section 4.2.) should be matched to the dimension and topographical shape of the survey area. Fig. 4 exemplifies how to place a track-net system in a lake and at an open coast with the aid of marking buoys.

Pilot survey; as part of the preparation work it is often beneficial to conduct a reconnaissance survey in order to identify obstacles, waterplants, piers etc that may be at hand. During the pilot study it is essential to

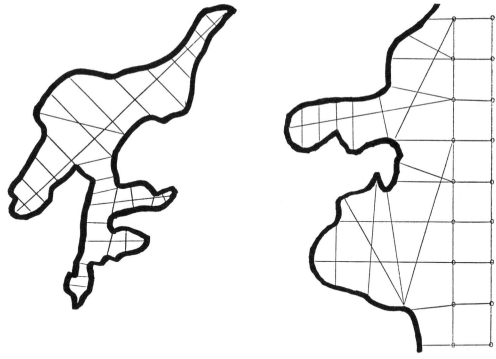

Fig. 4. One type of possible track-net system in a lake and an example of a
track-net system at an open coast with marking buoys

check out that the motor runs without problem, since it is a working principle and a necessity to keep constant speed during each course. The direction of the course, as well as the terminal points, should be marked on the
working map.

Calibration; before the field-work is initiated the echosounding equipment should be calibrated, to obtain relevant figures for the water depth
on the echogram. As a general rule it is advisable to calibrate the instru-
ment for different depth values against a lead-line sounder before, occasion-
ally during, and after the survey.

2.4. Field-work

A technique with "flying start" must always be practised (see Fig. 5). Two
persons are generally needed for the work. The boat-driver should keep a
constant course (compass) and a constant velocity (log). The operator should
keep a survey journal and mark on the echogram where the tracks start and
end, as well as deviations from the planned course. One type of survey journal is shown on Table 1.

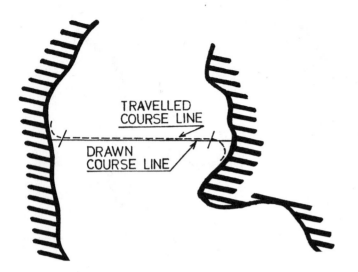

Fig. 5. "Flying start" of an echosounding track

Table 1. Survey journal

Area: **Lake Vänern**	Date: **771224**
Observer: **S. Klaus**	Assistant: **L. Håkanson**
Water level: **4441**	at gauge no: **5**
Water level:	at gauge no:
Sounding equipment no: **6**	

Control sounding	Time	Leadsounding	Echosounding
1	**10.00**	**12.0 m**	**12.0 m**
2	**18.00**	**25.0 m**	**25.4 m**

Depth scales: **0-50 m   I** ⎫
50-100 m   II ⎬ **a change indicated**
100-150 m   III ⎭ **with X on the echogram**

Start	Terminal points		Stop	Notations
10.10	**1 (30m)**	**2 (15m)**	**10.21**	
10.25	**2 (15m)**	**3 (18m)**	**10.38**	
10.45	**4 (50m)**	**5 (6m)**	**11.00**	**I and II**

3. BATHYMETRIC MAP CONSTRUCTION

The base for the bathymetric map construction is the echogram. After the field-work is finished the echogram for each track is cut into a separate strip. The depth intervals that are to be used on the bathymetric map should then be determined (see section 4.1. and 4.2.). The general principle is that the contour-line intervals (1_c in meters) should be positive integers. The relationship between contour-line interval (1_c) and number of contour lines (n) is given by:

$$\frac{D_{max}}{n} \sim 1_c \tag{1}$$

where D_{max} = the maximum recorded depth (in meters). For most scientific purposes, e.g. when bathymetric maps for whole lakes are concerned, one can conclude that between 4 - 20 contour lines are required. Formula (1) should be used in the following manner:
- the D_{max}-value in an arbitrary lake was found to be 56 m;
- the number of contour lines required (for the particular purpose of the survey) was determined to be n ⩾ 5. This gives:

$$\frac{56}{5} \geq 1_c \rightarrow 1_c \leq 11$$

Consequently, a contour-line interval of 10 m would be preferable in this hypothetical case, since this is an even figure which would provide a bathymetric map which is easy to read. It would yield a bathymetric map with the 1_c-values 10, 20, 30, 40 and 50 m. In general terms one may also conclude that an intense hydrographic survey (a small L_r-value according to formula 8) will provide material which allows for the construction of a bathymetric map with many contour lines; a less intense survey will yield a bathymetric map with fewer contour lines and hence also a large contour-line interval; this will imply a lower level of information.

The next step in the evaluation of the echograms is to mark all points corresponding to the chosen depth values (see Fig. 6). From these points parallel lines are drawn which cross the base line (the zero line) at right angles.

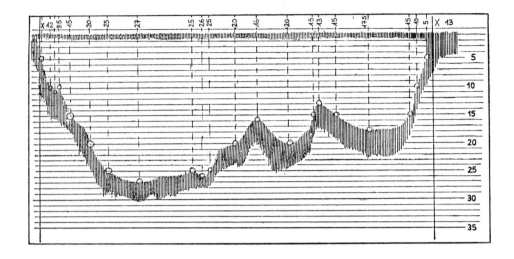

Fig. 6. Echogram with depth positions transferred to the base line

The next step is to transfer the information from the base line to the appropriate section on the map (see Fig. 7). The manual for this procedure is as follows:
- adjust all depth values to the given water level (the mean water level);
- place the starting point on the echogram on the corresponding starting point on the map section;
- turn the echogram an arbitrary angle (approximately 45°);
- draw a line between the ending points of the echogram and the map section (marked with x);
- transfer all recorded depth measurements and positions from the base line of the echogram to the map section by drawing lines parallel to the end- point line;
- mark the depth values on the map section and note the character (ups and downs) of the topography between the markings;
- draw contour lines on the bathymetric map by free hand. No mathematical or statistical interpolation method is generally needed.

Finally, it is often of great value to provide a complementary map illu- strating positions of the echosounded tracks. This will indicate the reliabi- lity of the bathymetric map. The information value of the bathymetric map is determined according to the method given in section 4.1.

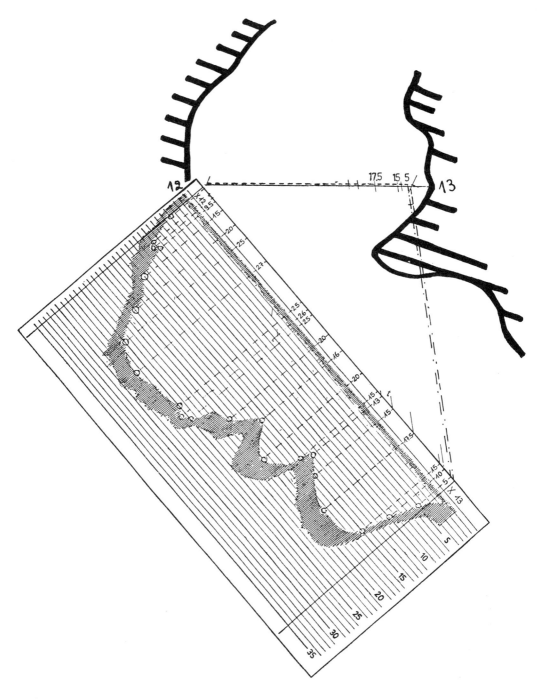

Fig. 7. Transformation of depth positions from the echogram to the map section

4. MORPHOMETRY

4.1. Optimization of lake hydrography surveys - the information value of bathymetric maps

Bathymetric maps are the source of most morphometric data used in limnological, hydrological and sedimentological contexts. The reliability of the morphometric data will depend on the accuracy of the bathymetric map, which in turn will depend on the intensity and accuracy of the hydrographic survey.

The aim of this part is to discuss the relationship between the intensity of the survey, aim of the survey and yield of the survey. A thorough analysis of this problem has been given by Håkanson (1978b).

The following functional relationships constitute the optimization model, which is built on the presumption that the survey is conducted with echo-sounding equipment along given tracks. The model is valid only for whole lakes:

$$I = I' \cdot I'' \tag{2}$$

$$I' = \frac{1}{a}\left[a - 0.14 \cdot L_r \cdot F^2 \cdot \sqrt{\frac{1}{n+2}} \cdot \sum_{i=1}^{n} \sqrt{a_i}\right] \tag{3}$$

$$I'' = \frac{e^{0.4 \cdot n} - 1}{e^{0.4 \cdot n} + 0.02} \tag{4}$$

This model is <u>not</u> as complicated as it first may seem.

The following parameters (which will be thoroughly discussed and defined in a subsequent section) are involved:

1. I = <u>the information value</u> of the bathymetric map. This is a figure between 0 and 1. It is 1 when the information in the bathymetric map is complete and completely correct.

2. I' = <u>correctly identified area</u> in the bathymetric map. The I'-value varies between 0 and 1. I' = 1 when all given contour lines are correctly placed. In that case we have an area error (E) equal to 0. The area error and the I'-value are defined in Fig. 8. We have:

$$I' = 1 - E \tag{5}$$

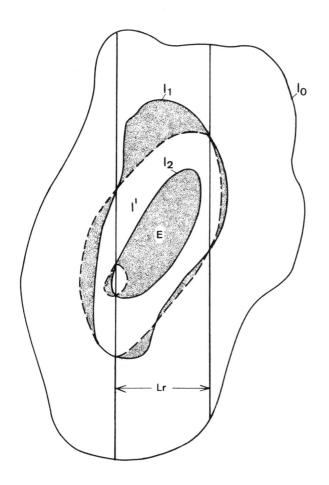

Fig. 8. Schematical illu-
stration of how
the area error
(E) is defined

——— real contour lines
– – – interpolated (estimated)
contour lines
▨▨▨ the area error (E)

3. I'' = the information number which depends on the number of contour lines
   (n) in the bathymetric map. The I''-value varies between 0 and 1 (see Table
   2). The I''-value, equation (4), describes the fact that a bathymetric map
   with few contour lines does not provide as good information about the bottom topography as a bathymetric map with many contour lines.

   The following interpretation code concerning the relationship between
   I, I' and I'' is valid for a given intensity of survey; the proportion of
   correctly identified area (I') on the bathymetric map increases as the
   number of contour lines (n) decrease, but as the number of contour lines

13

n	I''
1	0.3253
2	0.5458
3	0.6946
4	0.7949
5	0.8623
6	0.9076
7	0.9380
8	0.9585
9	0.9721
10	0.9813
12	0.9916
14	0.9962
16	0.9983
18	0.9992
20	0.9997
30	1.0000

Table 2. The relationship between the number of contour lines (n) and the information number (I'')

(n) decrease, there is a subsequent reduction in the information number (I''), and vice versa. The optimal information (I) will be obtained when the product $I' \cdot I''$ is at its maximum.

4. a = the lake area in km^2.

5. L_r = a measure of the intensity of the survey, e.g. the distance in km between parallel tracks (see Fig. 8). An L_r-value of 2 km will, for example, provide good material for map construction for large lakes ($a > 5\ 000$ km^2), but bad material for small lakes ($a < 2\ km^2$).

6. F = the normalized shore development (see eq. 24); a dimensionless parameter which indicates the degree of irregularity of the shoreline and of the lake bottom (Håkanson, 1974a). A lake with irregular topography must be surveyed with a greater intensity than a regular basin to obtain the same information value for the bathymetric map.

7. a_i = the total area (= the cumulative area) in km^2 within the limits of a given contour line (l_i). The term $\sum_{i=1}^{n} \sqrt{a_i}$ is directly related to the lake form concept (see section 4.4). Lakes with concave relative hypsographic curves must be more intensely surveyed than lakes with convex relative hypsographic curves to obtain the same information.

8. n = the number of contour lines in the bathymetric map; determines the I''-value.

9. e = the base for natural logarithms; $e = 2.718$.

The optimization model may be written in the following simplified form:

$$I = f(L_r, a, F, n, D) \tag{6}$$

That is, the information value (I) is governed by the intensity of the survey (L_r), the lake area (a), the topographical irregularity of the basin (F), the number of contour lines in the bathymetric map (n), and the lake form (D). This is also the order in which these variables influence the I-value.

4.2. The intensity of the survey

If the echosoundings are conducted along parallel tracks separated by a constant distance L_r, then the L_r-value may be used directly as a measure of the intensity of the survey. If the soundings are done in another manner, e.g. from a regular square net, then a representative value of the intensity of the survey may be determined from the following formula:

$$L_r = \frac{a}{s_n} \tag{7}$$

where a = the lake area in km^2;
       s_n = the total length of the echosounding tracks in km.

If the survey is conducted according to an irregular net, where the tracks are not parallel or crossing at right angles, then a representative L_r-value may be given from equation (7), if the tracks provide an even area cover of the whole lake surface.

If finally, the requirement for an even area cover is not at hand, then a representative L_r-value may be determined according to the following method.

Assume that an arbitrary lake has been surveyed according to a track net pattern illustrated on Fig. 9. The southern part of the lake is more intensely surveyed than the northern part. Many tracks are crossing. The requirement for even area cover is not met. The problem is to establish a representative L_r-value for the whole lake.

Assume that the lake area is 100 km^2 and that the total length of the echosounded tracks (s_n) is 55.0 km. Formula (7) gives an uncorrected L_r-value (L_r') of 100/55.0 = 1.82 km. The following method will yield a better L_r-value (see Fig. 10).

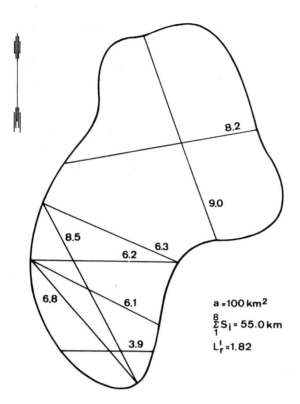

Fig. 9. Schematical illustration of 8 echosounded
tracks (length in km)
in an arbitrary lake

a = 100 km²

$\sum_{1}^{8} S_i = 55.0$ km

$L_r' = 1.82$

The longest of crossing tracks are used as reference tracks. There are
two such reference tracks in Fig. 10, 9.0 km and 8.5 km, labelled with a
ring. These tracks are fully counted. All tracks crossing these reference
tracks are reduced by a factor sin β, where β is the acute angle between the
tracks. The corrected L_r-value may then be calculated from the following formula:

$$L_r = \frac{a}{\sum_{i=1}^{n} s_i \cdot \sin \beta_i} \qquad (8)$$

The corrected L_r-value in this example is 2.30 km.

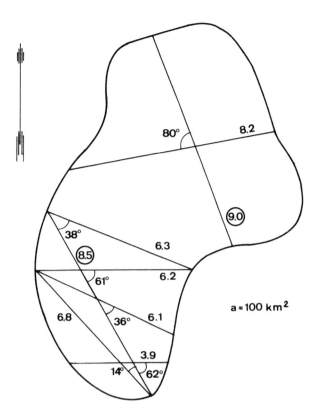

Fig. 10. Example illustrating a method to determine representative L_r-values

Within the figure: 80°, 8.2, 9.0, 38°, 6.3, 8.5, 61°, 6.2, 6.8, 36°, 6.1, a = 100 km², 3.9, 14°, 62°

s_i	β_i	$\sin \beta_i$	$s_i \cdot \sin \beta_i$
9.0	–	1	9.0
8.2	80	0.985	8.1
8.5	–	1	8.5
6.3	38	0.616	3.9
6.2	61	0.875	5.4
6.1	36	0.588	3.6
3.9	62	0.883	3.4
6.8	14	0.242	1.6

$\sum_1^8$ 55.0   43.5

$L_r' = 1.82$ $L_r = 2.30$

4.3. Practical use of the optimization model-manuals

In this section we will discuss two typical situations where the optimization model can be used. The first example concerns the manual for hydrographic surveys when no knowledge whatsoever is available about the bottom topography.

The only information at hand concerns lake area and shoreline length. The second example illustrates evaluation of a lake that has already been surveyed.

4.3.1. Manual - unknown lake

The aim is to determine an L_r-value.

Presuppositions: Lake area = 500 km^2;

Shoreline length = 400 km; normalized according to formula (15).

The shore development F is given by (see eq. 24):

$$F = \frac{400}{2 \cdot \sqrt{\pi} \cdot 500} = 5.05$$

- assume that we require a bathymetric map with an area error (E) that should be equal to or be less than 5 %, i.e. $I' \geqslant 0.95$;

- assume that we require a bathymetric map with at least 5 contour lines, i. e. $n \geqslant 5$;

These two requirements are given by the aim of the survey;

- assume that the lake form is $f(\bar{x})$.

The cumulative areas (a_i) for this hypothetical lake are given in Table 3, where $a_i = 500 \cdot a_i^P/100$. For $I' = 0.95$ we obtain (eq. 3):

$$0.95 = \frac{1}{500} \left[500 - 0.14 \cdot L_r \cdot 5.05^2 \cdot \sqrt{\frac{1}{5+2}} \cdot 61.0 \right]$$

That is, $L_r = 0.304$ km.

Irrespective of whether the field survey is done according to a deterministic or regular-net method, the L_r-value should be about 0.3 km. When the hydrographic survey has been carried out, a bathymetric map with 5 contour lines constructed (preferably in scale = 5 500 $\sqrt{a} \sim 123\ 000$, i.e. s $\sim 100\ 000$, see eq. 13), and a relative hypsographic curve drawn, then it may very well appear that the lake form is not $f(\bar{x})$ as assumed, but another type. Let us assume Lma, i.e. the relative hypsographic curve lies between the limitation lines given by $f(0.5)$ and $f(1.5)$, (see Fig. 11). Consequently, we must adjust the values according to the obtained empirical results. This may be done as follows:

18

Table 3. Cumulative areas for a hypothetical lake

n	a_i^P (%)	a_i (km^2)	$\sqrt{a_i}$ (km)
1	75.2	375.0	19.4
2	51.6	258.0	16.1
3	30.7	153.5	12.4
4	14.3	71.5	8.5
5	4.4	22.0	4.7
			$\sum\limits_1^5 \sqrt{a_i} = 61.0$

n	$a_i(e)$ (km^2)	$\sqrt{a_i}(e)$ (km)
1	448	21.2
2	391	19.8
3	315	17.7
4	215	14.7
5	99.0	9.9
		$\sum\limits_1^n \sqrt{a_i(e)} = 83.3$

where $a_i(e)$ = the empirically found cumulative areas that correspond to the given contour lines (see Fig. 11).

The corrected I'-value is then for $L_r = 0.30$ (eq. 3):

$$I' = 0.93$$

That is, a smaller yield than required. Three alternatives now exist. Either one can accept an I'-value of 0.93 for n = 5, or the initial requirement $I' \geqslant 0.95$ stands, then n must be changed. The third alternative is to change the L_r-value, i.e. to increase the field survey.

The required information value (I) is (eq. 2):

$$I = I' \cdot I'' = 0.95 \cdot 0.86 = 0.82$$

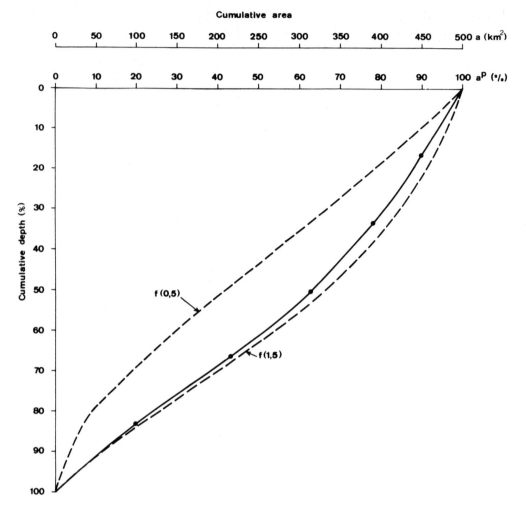

Cumulative area

<u>Fig. 11.</u> Hypothetical hypsographic curve for a lake of form Lma

The relationship between I, I$'$ and I$''$ for L_r = 0.30 km is illustrated in Table 4. The obtained I-value is:

I = 0.93 · 0.86 = 0.80

If the requirement for I$'$ ⩾ 0.95 stands, then n must be reduced. The question is - how much?

figure was arrived at from rather obscure premises. Nothing whatsoever was discussed in advance concerning the real aim of the survey. In the research program for the Lake Vänern project it was simply stated - "In this context, satisfactory information about the morphometry of Lake Vänern ought to be obtained with a net with a distance of 500 m between the tracks" (author´s translation from the Swedish). The total cost for this hypsographic survey, and the subsequent work to evaluate the echograms and construct a bathymetric map, was in the order of 500 000 - 1 000 000 Sw Kr (about 200 000 U.S. $).

We may focus on the Lake Vänern case simply because it is very typical in the sense that the survey was determined without any quantitative or qualitative analysis of intensity, aim and yield of the survey. In this respect, this particular survey, like most hypsographic surveys, may be considered very unscientific. This is quite remarkable, especially as hypsographic surveys are essential in practically all limnological contexts. They are fundamental for bathymetric maps, hypsographic curves, volume measurements, mean depth calculations etc. The costs and efforts involved are generally considerable.

The result of the hypsographic survey in Lake Vänern is given as a bathymetric map in scale 1:180 000 with 10 contour lines. A small scale copy of this new bathymetric map with only 5 contour lines is illustrated on Fig. 12.

The following questions concerning this new bathymetric map will now be discussed:
- what is the area error (E) of the bathymetric map in scale 1:180 000 with 10 contour lines, and how much of the area is correctly identified (the $I^{'}$-value)?
- what is the optimal information value (I), that may be obtained from the given intensity of the survey (L_r = 0.5 km)?
- what is the optimal number of contour lines, that can be used in the bathymetric map?

The area error (E) is determined accordingly:

Presuppositions: a = 5 648 km^2, L_r = 0.5 km, F = 7.14, n = 10,

$\sum\limits_{i=1}^{10} \sqrt{a_i}$ = 284.65 (see Table 5), i.e.:

from equation (3) we get:

$$I^{'} = \frac{1}{5\ 648} \left[5\ 648 - 0.14 \cdot 0.5 \cdot 7.14^2 \cdot \sqrt{\frac{1}{10+2}} \cdot 284.65 \right]$$
$$I^{'} = 0.948$$

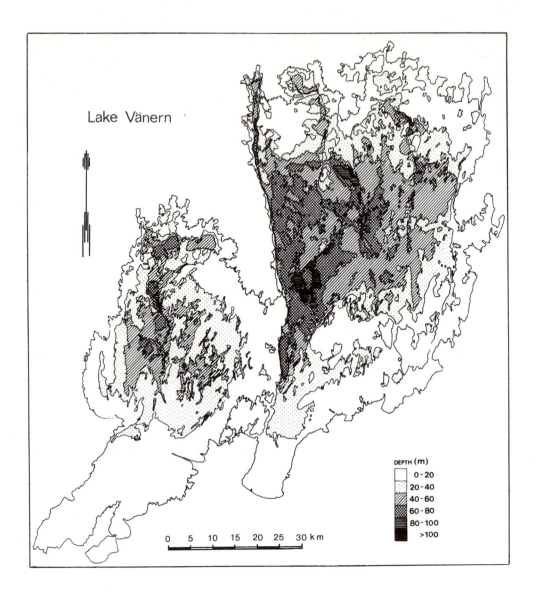

Fig. 12. Bathymetric map for Lake Vänern with 20 m contour-line interval, re-
drawn from the new bathymetric map in scale 1:180 000 with 10 contour lines

That is, the percentage of correctly identified area is 94.8. The area error
(E) is 5.2 % (100 - 94.8), or 294 km^2.

The total length of the contour lines in the bathymetric map is 11 400 km
(see later part). This implies that the average error, in length units, perpendicular to the contour lines is 0.026 km (294/11 400), or in scale
1:180 000, 0.14 mm. This is practically equal to the width of the contour

24

Table 5. Base data on depth and area in Lake Vänern

Depth (m)	Cumulative area (a_i) (km^2)	$\sqrt{a_i}$ (km)	n
0-10	5 648.02	75.15	0
10-20	3 995.18	63.21	1
20-30	2 957.46	54.38	2
30-40	2 233.67	47.26	3
40-50	1 603.86	40.05	4
50-60	1 025.84	32.03	5
60-70	544.87	23.34	6
70-80	208.95	14.46	7
80-90	48.94	7.00	8
90-100	5.36	2.32	9
100-106	0.37	0.61	10

$$\sum_{i=1}^{10} \sqrt{a_i} = 284.65$$

lines on the map. Thus, the error may, for most practical purposes, be regarded as negligible. A more intense survey may, however, change certain details in the bathymetric map, e.g. concerning new deep holes or sub-aquatic tops of limited areal extension, and the configuration of certain contour lines, which would be altered since the interpolation would be carried out from a new track net. Those hypothetical modifications can, however, only change details and not the general pattern of the contour lines in the bathymetric map of Lake Vänern.

The optimal information value $(I = I' \cdot I'')$ is determined in Table 6, which shows that the highest I-value is obtained for 13 contour lines (n = 13). This I-value is 0.935. However, it should be emphasized that the difference in I-values for n = 13 and n = 10 is very small indeed, I = 0.935, and I = 0.930, respectively. That is, in the third decimal. This implies that the new bathymetric map in scale 1:180 000 with 10 contour lines for Lake Vänern provides a very good information value for the given intensity of survey. A map with only 5 contour lines, like Fig. 12, gives a smaller area error (E) but also a comparatively low information number (I''). A map with 20 contour lines

Table 6. Determination of optimal information value (I) of, and optimal number of contour lines (n) in, the new bathymetric map of Lake Vänern

l_c (m)	n	I''	$\sum_1^n \sqrt{a_i}$ (km)	I'	I
50	2	0.546	32.64	0.990	0.541
30	3	0.696	72.92	0.979	0.682
25	4	0.795	93.97	0.976	0.776
20	5	0.862	125.38	0.970	0.836
15	7	0.938	178.26	0.949	0.890
12	8	0.958	232.63	0.954	0.913
10	10	0.981	284.65	0.948	0.930
9	11	0.987	320.63	0.944	0.932
8	13	0.994	366.37	0.940	0.935
6	17	0.999	499.64	0.928	0.927
5	21	1.000	606.41	0.920	0.920
4	26	1.000	767.99	0.908	0.908

will give a higher information value (I'') but a comparatively high area error (E). Real topographical features vanish if too few contour lines are utilized in the map. Irrelevant topographical elements appear if too many contour lines are used.

4.4. Morphometrical parameters

Maximum length (L_{max} in km); defined by the line connecting the two most remote points on the shoreline. In regular basins this line is generally straight and concurs with the maximum effective length (L_e). In irregular lakes, e.g. oxbow lakes, L_{max} is a curved line. It may not cross land, but it may cross islands. Consequently, the maximum length cannot always be given a definite value. It has limited limnological use, and is primarily to be considered as a descriptive measure.

The maximum length is 140 km in Lake Vänern (see Fig. 13).

Maximum effective length (L_e in km); defined by the straight line connecting the two most distant points on the shoreline over which wind and waves may act without interruptions from land or islands. This is an important parameter in many limnological and hydrological contexts, e.g. concerning internal seiches (see Fig. 14).

The maximum effective length is 97 km in Lake Vänern (Fig. 13).

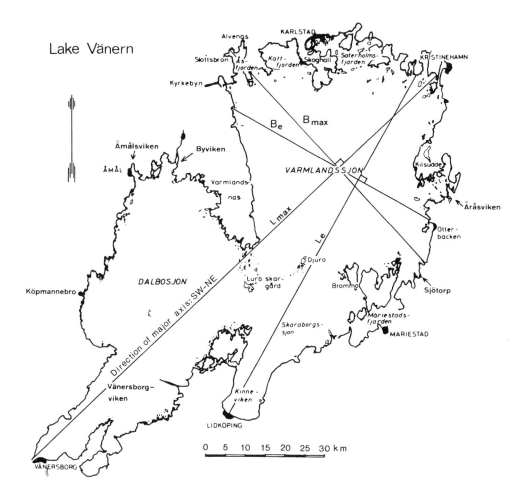

Fig. 13. Orientation map for Lake Vänern, and illustrations of the maximum
length (L_{max}), the maximum effective length (L_e), the maximum width
(B_{max}) and the maximum effective width (B_e)

Effective length (L_s in km); defined by the straight line from an arbitrary
position on the lake to the most distant point on the shoreline without crossing land or islands, which may reduce the impact of wind-induced waves.

Effective fetch (L_f in km); defined according to a method introduced by
Beach Erosion Board (1972) (Fig. 15). The effective fetch gives a more representative measure of how the wind governs the waves (wave length, wave height)
than the effective length, since several wind directions are taken into account.
A proper figure of the effective fetch is rather easy to determine by means
of a special transparent paper illustrated on Fig. 16. The central radial of

27

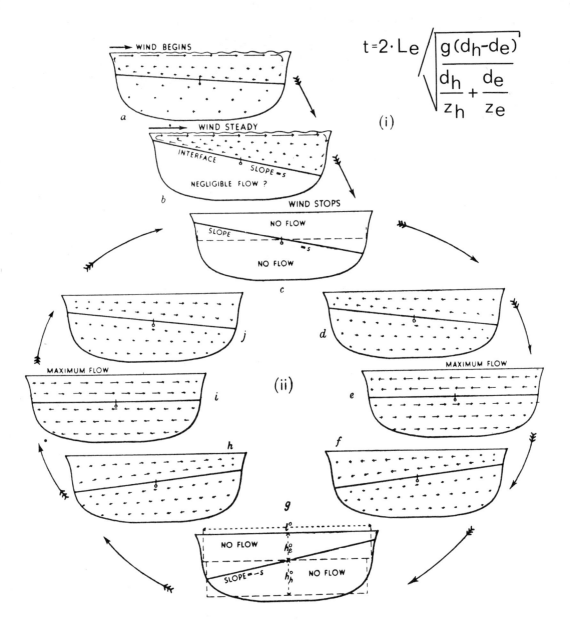

Fig. 14. Movement caused by (i) wind stress and (ii) subsequent interval seiche in a hypothetical two-layered lake, neglecting friction. Direction and velocity of flow are approximately indicated by arrows. σ = model section; t = period of the internal seiche; L_e = maximum effective length; g = acceleration due to gravity (980.6 cm/sec); d_e = density of epilimnion water; d_h = density of hypolimnion water; z_e = thickness of homogeneous epilimnion; z_h = thickness of hypolimnion (from Mortimer, 1952 and Wetzel, 1975)

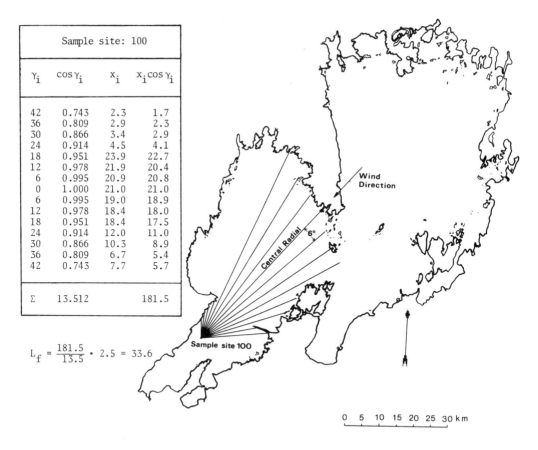

γ_i	cos γ_i	x_i	x_i cos γ_i
42	0.743	2.3	1.7
36	0.809	2.9	2.3
30	0.866	3.4	2.9
24	0.914	4.5	4.1
18	0.951	23.9	22.7
12	0.978	21.9	20.4
6	0.995	20.9	20.8
0	1.000	21.0	21.0
6	0.995	19.0	18.9
12	0.978	18.4	18.0
18	0.951	18.4	17.5
24	0.914	12.0	11.0
30	0.866	10.3	8.9
36	0.809	6.7	5.4
42	0.743	7.7	5.7
Σ	13.512		181.5

Sample site: 100

$$L_f = \frac{181.5}{13.5} \cdot 2.5 = 33.6$$

Fig. 15. Determination of effective fetch (L_f). Example from sample site 100 in Lake Vänern

this transparent paper is simply put in the main wind direction or, if the maximum effective fetch is required, in that direction which gives the highest L_f-value. Then the distance (x in km) from the given station, or sample site, to land, or to island, is measured for every deviation angle γ_i, where $\gamma_i = \pm 6°$, $\pm 12°$, $\pm 18°$, $\pm 24°$, $\pm 30°$, $\pm 36°$, and $\pm 42°$. The effective fetch may subsequently be calculated from the formula:

$$L_f = \frac{\Sigma x_i \cos \gamma_i}{\Sigma \cos \gamma_i} \cdot S \tag{9}$$

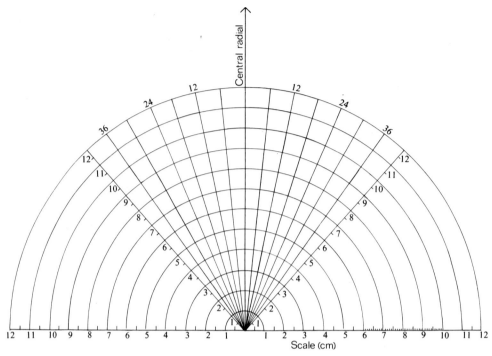

Fig. 16. Diagram illustrating a simplified technique for establishing the effective fetch (L_f). The diagram should be on transparent paper

where $\Sigma \cos \gamma_i$ = 13.5, a constant;

S' = the scale constant; S' = 2.5 for the map scale 1:250 000, and 1.0 for the scale 1:100 000.

The areal distribution of the maximum effective fetch in Lake Vänern is illustrated on Fig. 17.

The effective fetch is an important parameter in, e.g. beach morphological contexts (Fig. 18) and to establish the prevailing bottom dynamic situation (erosion-transportation-accumulation), (see Fig. 19).

Maximum width (B_{max} in km); defined by the straight line at a right angle to the maximum length (L_{max}), which connects the two most remote extremities on the shoreline without crossing land. Islands may be crossed. B_{max} has, like L_{max}, primarily a descriptive value in limnological contexts.

B_{max} = 72 km in Lake Vänern (Fig. 13).

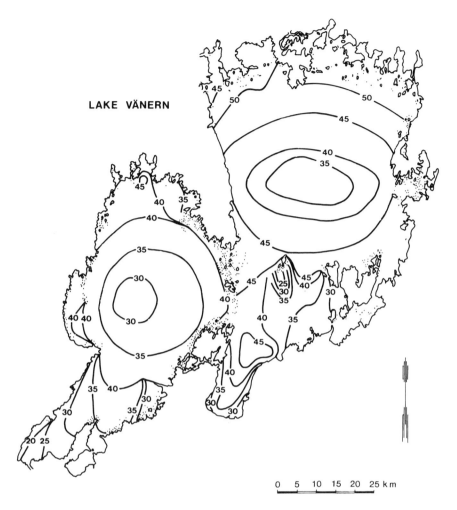

LAKE VÄNERN

0   5   10   15   20   25 km

Fig. 17. The areal distribution of the effective fetch (L_f) in Lake Vänern

Maximum effective width (B_e in km); defined by the straight line on the lake surface, perpendicular to the maximum effective length (L_e), which connects the two most distant points on the shoreline. Thus, B_e may not cross land or islands.

B_e = 57 km in Lake Vänern (Fig. 13).

Mean width ($\bar{B}$ in km); defined by the ratio lake area (a in km^2) to maximum length (L_{max} in km), i.e.:

$$\bar{B} = \frac{a}{L_{max}}$$
(10)

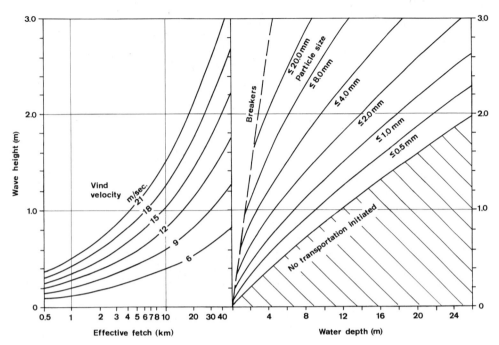

Fig. 18. The relationship between effective fetch (L_f) and wave height and between wave height and bottom dynamics in the shore zone (after Norrman, 1964)

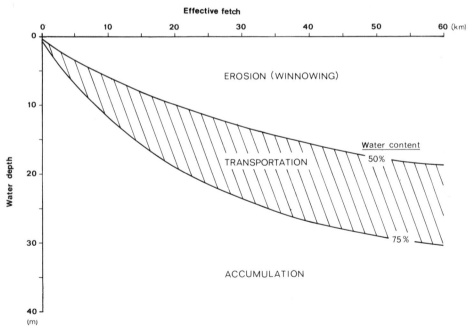

Fig. 19. The ETA-diagram (erosion-transportation-accumulation) for surficial sediments (0 - 1 cm). From Håkanson (1977b)

The maximum effective width and the mean width are, like the maximum effective length, valuable hydromechanical tools.

The mean width is 40 km in Lake Vänern.

Maximum depth (D_{max} in m); the greatest known depth of the lake.

D_{max} = 106 m in Lake Vänern.

Mean depth ($\bar{D}$ in m); defined by the quotient lake volume (V in km^3) to lake area (a in km^2), i.e.:

$$\bar{D} = \frac{1\ 000 \cdot V}{a} \tag{11}$$

The $\bar{D}$-value is a most useful parameter, e.g. in models describing the productivity and the trophic status of lakes (Figs. 20 and 21).

The mean depth is 27 m in Lake Vänern.

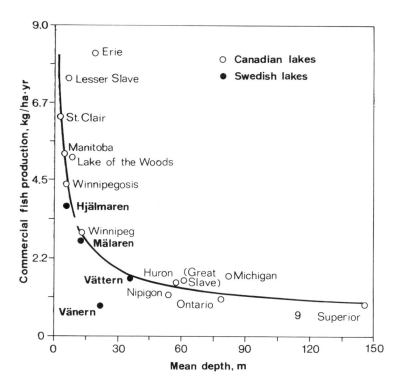

Fig. 20. The relationship between commercial fish production and mean depth (from Rawson, 1955 and Ahl, 1975)

33

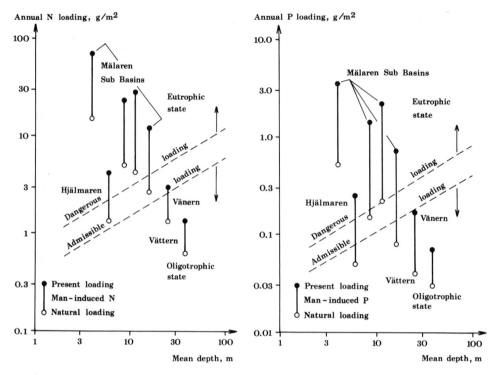

Fig. 21. The loading of N and P in relation to trophic level and mean depth
in the Swedish Great Lakes (from Vollenweider, 1968 and Ahl, 1975)

Median depth (D_{50} in m); according to definition, 50 % of the lake area
should lie below the D_{50}-value and 50 % above. The median depth is determined from the percentage hypsographic curve (or the hypsographic curve) (see
Fig. 22).

D_{50} = 22 m in Lake Vänern.

In lakes with convex relative hypsographic curves, like Lake Vänern, the
mean depth ($\bar{D}$) is larger than the median length (D_{50}). In lakes with concave
relative hypsographic curves the opposite is valid.

The median depth may be used, e.g. to determine the lake bottom roughness
(R), which is a useful parameter in sedimentological contexts and in the optimization model for lake hydrographic surveys.

Quartile depths (D_{25} and D_{75}); according to definition, 25 % of the lake
area should lie below and 75 % above the depth given by the first quartile
depth value (D_{25}). In the same way we have 75 % of the lake area below and
25 % of the lake area above the D_{75}-value. The median depth (D_{50}) may also
be called the second quartile depth. The quartile depths are primarily to be

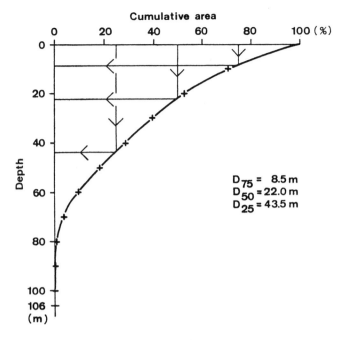

Fig. 22. Determination of the median depth (D_{50}) and the quartile depths
         (D_{25}, D_{75}) from the percentage hypsographic curves for Lake Vänern

considered as descriptive morphometrical parameters, which are determined
from the percentage hypsographic curve (Fig. 22).

   D_{25} = 43.5 m and D_{75} = 8.5 m in Lake Vänern.

Relative depth (D_r in %); defined by the ratio of maximum depth (D_{max} in m)
to mean diameter of the lake, i.e.:

$$D_r = \frac{D_{max} \cdot \sqrt{\pi}}{20 \cdot \sqrt{a}}$$ (12)

where a = the lake area in km^2.

   The relative depth is 0.12 % in Lake Vänern, which is a normal figure for
large basins. Small and deep lakes have high D_r-values. The relative depth
may be used to describe stability of stratification of lakes. Eberly (1964)
states that "Morphometry (both area and relative depth) appears to be a sig-
nificant factor in the development of the metalimnetic oxygen maximum in the
U.S.-lakes."

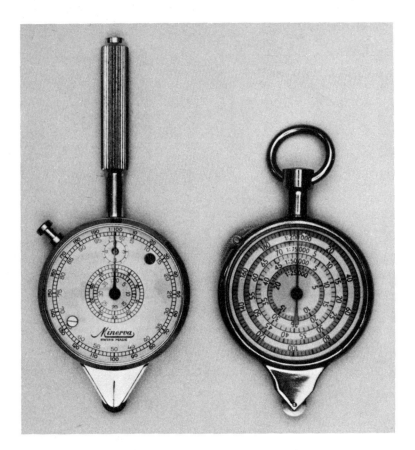

Fig. 23. Two types of map measurer

Direction of major axis; defined by the general compass direction of the maximum length (L_{max}). It is SW - NE in Lake Vänern.

Shoreline length (l_o in km); can be determined with map measurer (Fig. 23), or better with CTP-technique (= checkered transparent paper), (see Håkanson, 1978a).

The manual for determinations of shoreline length with CTP-technique is accordingly:

1. The total lake area (A), i.e. the lake area plus the total area of islands, is assumed to be known. A = 5 893 km for Lake Vänern.

2. The optimal map scale for the determination of the shoreline length is given by:

$$s \sim 5\ 500 \cdot \sqrt{A} \tag{13}$$

36

where s = the scale factor, equal to 1/S, where S is the map scale.

That is, for Lake Vänern:

$$s \sim 5\ 500 \cdot \sqrt{5\ 893} \sim 422\ 000$$

The best, i.e. the largest, available map scale that corresponds to this scale factor value should be used. For Lake Vänern there exist possible maps in scales 1:250 000 (best) and 1:500 000.

3. The checkered transparent paper, which has a square size of 0.5 cm, is then put over the map and the number of intersectional points (x) between the shoreline and the squares is counted. The method is illustrated on Fig. 24. From the map in scale 1:250 000 a total of 1 286 intersectional points were obtained. No further determination is then required to establish the shoreline length with a given statistical confidence, i.e. the error in the length determination should, with a 95 % certainty, be less than 5 %.

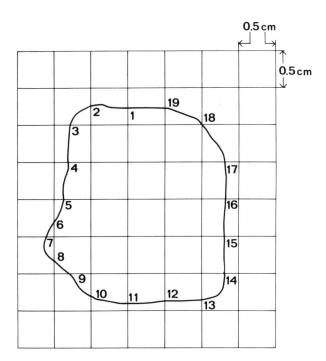

Fig. 24. Schematical illustration of CTP-technique for determinations of the shoreline length (l_0). The number of intersectional points in this example is x = 19

If, on the other hand, we had chosen another map scale, e.g. 1:1 000 000, then we would have a different x-value. Let us say x = 200. Thus to obtain the shoreline length with the same statistical certainty, we must make more determinations. The relationship between the number of necessary turns or determinations (n'), i.e. the number of times the checkered transparent paper is put on the map and the x-value determined, and the number of intersectional points (x) is given in Table 7. From this table we get that n' = 5 for x = 200, i.e. four more determinations have to be done to obtain the shoreline with the given statistical certainty. The mean value ($\bar{x}$) of the 5 measurements is 202.

$\bar{x}$	n'
> 475	1
474 - 350	2
349 - 260	3
259 - 205	4
204 - 160	5
159 - 125	6
124 - 100	7
99 - 82	8
81 - 70	9
69 - 59	10
58 - 51	11
50 - 46	12
45 - 42	13
41 - 39	14
38 - 36	15
35 - 34	16
33	17
32 - 31	18
30	19
29	20

Table 7. The relationship between the number of necessary turns (n'), to obtain the shoreline length with a given statistical certainty, and the number of intersectional points ($\bar{x}$) obtained with CTP-technique

4. The scale dependent shoreline length ($1'_o$) can then be established from the following formula:

$$1'_o = 0.5 \cdot \bar{x} \cdot s \cdot 10^{-5} \qquad (14)$$

That is, for Lake Vänern from the map in scale 1:250 000 we get:

$$1'_o = 0.5 \cdot 1\ 286 \cdot 250\ 000 \cdot 10^{-5}$$
$$1'_o = 1\ 608\ km$$

and from the map in scale 1:1 000 000:

$$1_o' = 0.5 \cdot 202 \cdot 1\ 000\ 000 \cdot 10^{-5}$$

$$1_o' = 1\ 012 \text{ km}$$

Thus, the shoreline length depends on the map scale.

5. The requested normalized scale independent value of the shoreline length (1_o) may then be calculated from the normalization formula:

$$1_o = \frac{1_o' \cdot (K_2 - K_1)}{K_2 - \log(s + a')} \tag{15}$$

where $a' = 10^5 \cdot \log A$; 10^5 = the area constant, A = the total lake area in km^2;

     s = the scale factor (= 1/S);

     $K_1 = \log(s + a')$ for s = 1, i.e. the reference scale (1:1);

     $K_2 = \log(s + a')$ for s = 6 000 000, where 6 000 000 = scale constant.

The data from the map in scale 1:250 000 for Lake Vänern gives:

$$1_o = \frac{1\ 608 \cdot \left[\log(6\ 000\ 000 + 10^5 \cdot \log 5\ 893) - \log(1 + 10^5 \cdot \log 5\ 893)\right]}{\log(6\ 000\ 000 + 10^5 \cdot \log 5\ 893) - \log(250\ 000 + 10^5 \cdot \log 5\ 893)}$$

$$1_o = 1\ 960 \text{ km}$$

The data from the map in scale 1:1 000 000 gives, in the same way, 1_o = 1 870 km. The difference between those two determinations is 90 km, or 4.6 %, which is within the acceptable statistical limits. Since the figure 1 960 km is obtained from a better map, this value is preferable.

The mean value from numerous determinations of the shoreline length of Lake Vänern is 1 943 km (see Håkanson, 1978a).

To determine the shoreline length is important, e.g. to establish the shore development (F) and to calculate the production of benthic algae and higher water vegetation.

<u>Contour-line length</u> $(1_i$ in km); a normalized scale independent value of an arbitrary contour line (1_i) may be determined with CTP-technique in the same manner as illustrated for the shoreline length (1_o). The area $(a_i$ in $km^2)$ should in such cases be given by the total area within the limits of the actual contour line.

The normalized length of any contour line in a bathymetric map may also be estimated from the following formula (Håkanson, 1978a):

$$l_i^+ = 2.1 \cdot F^{3/2} \cdot \sqrt{a_i} \qquad (16)$$

where l_i^+ = the scale independent estimated length of the contour line in km;

   2.1 = an empirical constant;

   F = the normalized shore development, dimensionless;

   a_i = the total area within the limits of the given contour line in km^2.

The total length (normalized) of all contour lines in a bathymetric map (L^+ in km) may consequently be estimated as follows:

$$L^+ = \sum_{i=1}^{n} \cdot l_i^+ = 2.1 \cdot F^{3/2} \cdot \sum_{i=1}^{n} \sqrt{a_i} \qquad (17)$$

The total length (L^+) of all the contour lines in the bathymetric map in scale 1:180 000 with 10 m contour line interval for Lake Vänern is 11 400 km (the shoreline excluded).

The L^+-value is, for example, utilized to determine the lake bottom roughness (R).

Total lake area (A in km^2); i.e. the lake area (a) plus the area of all islands, islets and rocks within the limits of the shoreline. This area value is generally determined with a planimeter (Fig. 25), (see also Welch, 1948), from a map in given scale. If the scale factor (s) is increased small capes and bays will be cancelled out on the map. This will primarily affect determinations of the shoreline length and not the area determination, since planimetration implies that negative errors (bays) are levelled out by positive errors (capes); and provided the planimetration is carried out on a map of reasonable scale, which is given by the relationship $s \leqslant 5\ 500\ \sqrt{A}$, one can regard the area determination as rather independent of the map scale.

A = 5 893 km^2 in Lake Vänern.

The A-value is used to determine the shore development (F), which is an important parameter in the optimization model for hydrographic surveys.

The A-value is also closely related to the insulosity concept.

Lake area (a in km^2); i.e. the water surface. a = 5 648 km^2 in Lake Vänern.

Fig. 25. The ARISTO planimeter

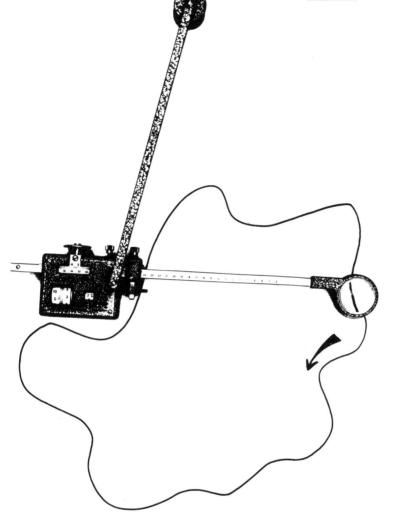

<u>Volume</u> (V _ km^3); may be determined from the following two formulae:

$$V_1 = \sum_{i=0}^{n} \frac{1_c}{2} (a_i + a_{i+1})$$ (18)

$$V_p = \sum_{i=0}^{n} \frac{1_c}{3} (a_i + a_{i+1} + \sqrt{a_i \cdot a_{i+1}})$$ (19)

where 1_c = the contour-line interval in m;

a_i = the total area (= the cumulative area) within the limits of the
contour line 1_i, in km^2.

The V_1-formula (eq. 18) is called the linear approximation of the volume, the V_p-formula (eq. 19) the parabolic approximation. Which of these two formulae should be used depends on the lake form (Håkanson, 1977a). The V_1-formula provides best values for lakes with concave relative hypsographic curves, the V_p-formula is preferable for lakes with convex relative hypsographic curves. The manual for lake volume determination will now be exemplified for Lake Vänern. Base data on depth and area values are given in Table 8. Lake Vänern has a form of the type defined as convex macro (Cxma), i.e. the relative hypsographic curve lies between the limitation lines $f(-1.5)$ and $f(-0.5)$ and it has no point of inflexion (see Fig. 26).

Table 8. Base data on depth and area in Lake Vänern

Depth (m)	Cumulative depth (%)	Area (km^2)	Cumulative area (km^2)	Cumulative area (%)
0 - 10	0	1 652.84	5 648.02	100
10 - 20	9.43	1 037.72	3 995.18	70.7
20 - 30	18.9	723.74	2 957.46	52.4
30 - 40	28.3	629.81	2 233.67	39.5
40 - 50	37.7	578.02	1 603.86	28.4
50 - 60	47.2	480.97	1 025.84	18.2
60 - 70	56.6	335.92	544.87	9.65
70 - 80	66.0	160.01	208.95	3.70
80 - 90	75.5	43.58	48.94	0.87
90 - 100	84.9	4.99	5.36	0.09
100 - 106	94.3	0.37	0.37	0.01
106	100	0	0	0

The uncorrected volume, according to the V_p-formula, is 153.45 km^3. Now, how can this value be corrected and what is the certainty of the volume determination?

A correction factor is obtained from Table 9. With 10 contour lines and a convex (Cx) lake form we get from this table a correction factor of 0.996 for Lake Vänern. That is, the corrected volume is 0.996 · 153.45 = 152.84 km^3. The true volume must lie between the limits obtained when the correction factor for the limitation lines $f(-1.5)$ and $f(-0.5)$ are multiplied with the uncorrected V_p-value of 153.45 km^3. The correction factor for the $f(-1.5)$-line, for n = 10, is 0.9896, and the correction factor for the $f(-0.5)$-line, for

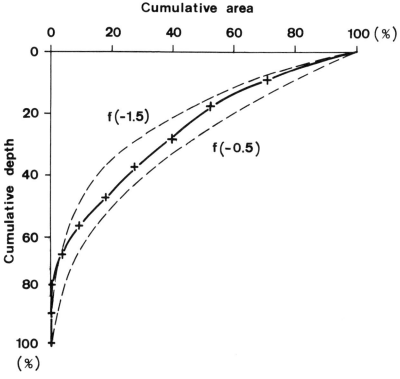

The relative hypsographic curve for Lake Vänern between the limitation lines f(-1.5) and f(-0.5)

Table 9. Correction factor table for various lake forms and various numbers of contour lines in the bathymetric map (n)

		n=2	n=3	n=4	n=5	n=6	n=8	n=10
V_p	VCX	0.627	0.799	0.878	0.916	0.945	0.966	0.977
	Cx	0.930	0.969	0.981	0.988	0.990	0.993	0.996
	SCx	1.033	1.017	1.010	1.005	1.004	1.003	1.002
V_l	L	0.984	0.992	0.997	0.999	0.999	1.000	1.000
	C	1.046	1.023	1.012	1.007	1.006	1.003	1.002

n = 10, is 0.9993 (see Table 10). That is, the real volume must lie between 153.39 km^3 (= 0.9993 · 153.45) and 151.85 km^3 (= 0.9896 · 153.45). The maximum percentage error in the corrected volume 152.84 km^3 is between +0.4 % to -0.6 %.

Table. 10. Correction factor table for various defined relative hypsographic curves in the classification system for lake forms

n	f(-3)	f(-2)	f(-1.5)	f(-1)	f(-0.5)	f($\bar{x}$)
			V_p			
1	0	0	0.2704	0.6609	0.9147	1.1068
2	0	0.5391	0.7883	0.9124	0.9889	1.0328
5	0.3581	0.8927	0.9595	0.9847	0.9974	1.0047
10	0.7468	0.9709	0.9896	0.9944	0.9993	1.0017
25	0.9435	0.9938	0.9990	0.9996	1.0000	1.0001
50	0.9952	0.9986	0.9997	1.0000	1.0000	1.0000

n	f($\bar{x}$)	f(0.5)	f(1)	f(1.5)	f(2)	f(3)
			V_1			
1	0.6601	0.8654	1.0208	1.1563	1.2610	1.4037
2	0.9200	0.9585	0.9911	1.0213	1.0593	1.1354
5	0.9865	0.9966	0.9966	1.0020	1.0098	1.0357
10	0.9972	0.9989	0.9998	1.0002	1.0024	1.0113
25	0.9993	0.9998	1.0000	1.0000	1.0003	1.0023
50	0.9999	1.0000	1.0000	1.0000	1.0000	1.0004

This method to determine the lake volume is, in principle, valid for all lake types. However, for lakes with rather more complex relative hypsographic curves the correction factor has to be established in a different manner. This procedure will now be illustrated by using data from another Swedish lake basin, namely Great-Hjälmaren. Data on depth and area are given in Table 11; the relative hypsographic curve is illustrated on Fig. 27. There are 7 contour lines (n=7) in this particular case. Fig. 27 shows that 28 % of the area given by the relative hypsographic curve falls within the limits that signify convex (Cx) lake form, i.e. between the lines called f(-1.5) and f(-0.5); 72 % of the area emanates from the SCx-class given by the lines f(-0.5) and f(0.5). Thus, the SCx-class is the main class and this implies that the V_p-formula (eq. 19) should be utilized in the volume determination.

Table 11. Base data on depth and area in Great-Hjälmaren

Depth (m)	Cum. depth (%)	Area (km^2)	Cum. area (km^2)	Cum. area (%)	n
0-3	0	56.15	277.47	100	-
3-6	13.6	44.36	221.32	80.0	1
6-9	27.3	88.55	176.96	64.0	2
9-12	40.9	54.88	88.42	32.0	3
12-15	54.5	19.67	33.54	12.1	4
15-18	68.2	12.16	13.87	5.0	5
18-21	81.8	1.69	1.71	0.62	6
21-22	95.5	0.02	0.02	0.007	7
22	100	0	0	0	-

The uncorrected value of the volume of Great-Hjälmaren (V_p') may then be calculated as 2.0010 km^3. Subsequently, a correction factor may be established by using data from Table 9. For the Cx-class we obtain for n=7 a value of (0.990 + 0.993)/2 = 0.9915, and for the SCx-class we get (1.003 + 1.004)/2 = 1.0035. A normalized correction factor may then be determined accordingly:

$$k = a_{SCx} \cdot k_{SCx} + a_{Cx} \cdot k_{Cx} \tag{20}$$

where k = the normalized correction factor;

a_{SCx} = the area within the SCx-class, i.e. 72 %;

k_{SCx} = the correction factor for the SCx-class, i.e. 1.0035;

a_{Cx} = the area within the Cx-class, i.e. 28 %;

k_{Cx} = the correction factor for the Cx-class, i.e. 0.9915.

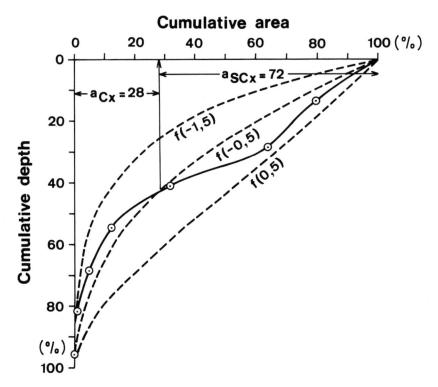

Fig. 27. The relative hypsographic curve for Great-Hjälmaren

That is:

$$k = \frac{72}{100} \cdot 1.0035 + \frac{28}{100} \cdot 0.9915 = 1.00014$$

The corrected volume (V_p) of Great-Hjälmaren is then:

$$V_p = V'_p \cdot k = 2.0010 \cdot 1.00014 = 2.0013 \text{ km}^3$$

Slope (α); the slope for an arbitrary station in a lake may be determined directly from the echogram, provided the track follows the same direction as the major slope axis, which can be determined from the bathymetric map. The slope is given in degrees (α^0) or preferably as a percentage (α_p) of the height versus the length (see Fig. 28). The slope between two contour lines in the bathymetric map can be determined from the formula (see Fig. 29):

$$\alpha_p = \frac{(l_1 + l_2) \cdot l_c}{20 \cdot a''} \tag{21}$$

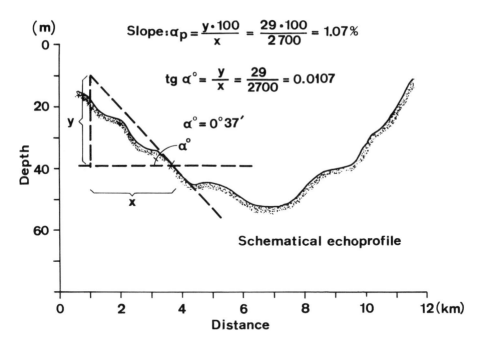

$$\text{Slope:}\,\alpha_p = \frac{y \cdot 100}{x} = \frac{29 \cdot 100}{2\,700} = 1.07\%$$

$$\text{tg}\,\alpha^\circ = \frac{y}{x} = \frac{29}{2700} = 0.0107$$

$$\alpha^\circ = 0^\circ 37'$$

Schematical echoprofile

Fig. 28. Example illustrating the determination of the slope, α_p in percent, and α° in degrees, from an echogram

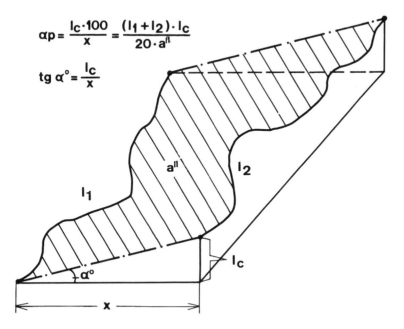

$$\alpha_p = \frac{l_c \cdot 100}{x} = \frac{(l_1 + l_2) \cdot l_c}{20 \cdot a^{fl}}$$

$$\text{tg}\,\alpha^\circ = \frac{l_c}{x}$$

Fig. 29. Schematical illustration of the determination of the slope between two contour lines (l_1 and l_2) in the bathymetric map

where α_p = the slope in %;

l_1, l_2 = the length of the two contour lines in km;

l_c = the contour-line interval in m;

a" = the area between the two contour lines in km^2. a" = $a_1 - a_2$, where a_1 and a_2 = the cumulative area (in km^2) limited by the two contour lines.

The slope is an important morphometrical parameter in many sedimentological contexts, e.g. in connexion with slumping and turbidity currents (see e.g. Gilbert, 1975 and Sturm, 1975), (Table 12).

Fig. 30 illustrates the relationship between the slope (α_p), determined from echograms, and the physical status of the surficial sediments (0 - 1 cm), as this is represented by the water content (W_{0-1}), within a test area in central Lake Vänern. The position of this test area is such that one would expect uniform sedimentological conditions (accumulation area). The major factor controlling the physical sediment character within this area is the slope. No fine material (medium silt and finer) is deposited on these slopes if the inclination is larger than 4.6 %.

Mean slope ($\bar{\alpha}$); defined for the entire lake area. It may be determined from the formula:

$$\bar{\alpha}_p = \frac{(l_0/2 + l_1 + l_2 + \cdots\cdots + l_{n-1} + l_n/2) \cdot D_{max}}{10 \cdot n \cdot a} \quad (22)$$

where $\bar{\alpha}_p$ = the mean slope in %;

D_{max} = the maximum depth in m;

l_i, i = 1, 2,...., n = the normalized length of the contour lines in km;

l_0 = the normalized shoreline length in km;

n = the number of contour lines;

a = the lake area in km^2.

Since the length of the last contour line (l_n) is generally small compared to the total length of the rest of the contour lines, formula (22) may be simplified accordingly:

$$\bar{\alpha}_p = \frac{(l_0 + 2 \cdot l) \cdot D_{max}}{20 \cdot n \cdot a} \quad (23)$$

Table 12. Depositional environments in Lake Brienz (from Sturm, 1975)

	DELTA AREA	CENTRAL BASIN PLAIN	LATERAL SLOPE
AREAL DISTRIBUTION (in % of lake floor)	38 %	23%	40 %
RELIEF	rough, uneven	smooth, flat	smooth, steep
INCLINATION	8°- 20°(foreset)	0°- 1°	30°- 40°
SEDIMENT SOURCE	proximal	proximal ▸ ◈ ◈ distal	distal
ENERGY LEVEL	high	low, occ. high	low
ENVIRONMENT	erosional = depositional	erosional ≪ depositional	depositional
MODE OF TRANSPORT	turbidity current	turbidity curr./ undercurrent	undercurrent
CHANNELS	common	not distinct	absent
SEDIMENTATION RATE (last 75 years)	40 - 50 mm/y	5.5 - 6.8 mm/y	2.5 - 3.3 mm/y
GAS EXPANSION HOLES	common	common	± absent
GRAIN SIZE DISTRIBUTION			
CLAY/SILT/SAND - RATIO			
SEDIMENTARY STRUCTURES			

49

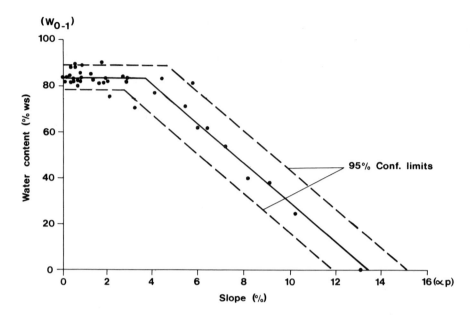

Fig. 30. The relationship between slope (α_p) and water content of surficial
sediments, 0 - 1 cm (W_{0-1}), within a test area in central Lake Vänern
(from Håkanson, 1977b)

where $1 = \sum\limits_{i=1}^{n} 1_i$; i.e. 1 = the total normalized length of the contour lines

in the bathymetric map in km. This value may be estimated from
equation (17), i.e. by the L^{+}-value;

α_p = 2.32 % in Lake Vänern.

Median slope (α_{50}); can be determined from the slope curve (Fig. 31). The median slope (α_{50}) signifies that 50 % of the lake area inclines more than the
α_{50}-value and 50 % less than this value.

The median slope is 0.75 % in Lake Vänern. The mean slope and the median
slope are primarily to be considered as descriptive parameters.

Shore development (F, dimensionless); a measure of the degree of irregularity of the shoreline. The F-value is defined by the formula:

$$F = \frac{1_o}{2 \cdot \sqrt{\pi A}}$$
(24)

where 1_o = the normalized shoreline length in km;
A = the total lake area in km^2.

50

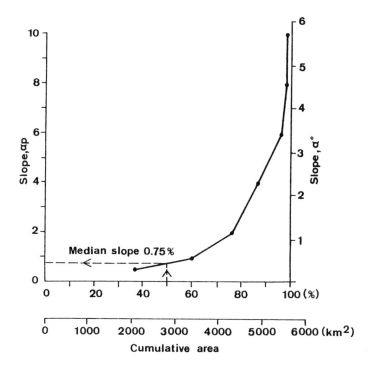

Fig. 31. Determination of the median slope (α_{50}) from the slope curve for Lake Vänern

The F-value illustrates the relationship between the actual length of the shoreline and the length of the circumference of a circle with an area equal to the total lake area. A perfect circular basin has a F-value of 1. Irregular lakes have higher F-values. F-values larger than 10 are rare. Lake Mälaren, a lake which probably has one of the most irregular basins in the world, it is practically an archipelago in itself, has a normalized F-value of 9.95.

The shore development is a key parameter in the optimization model for lake hydrographic surveys. It is highly correlated to the lake bottom roughness (R); the correlation coefficient is 0.98 (Håkanson, 1974a), which is quite understandable for morphological reasons; a lake with an irregular bottom will also have an irregular shoreline.

It should be emphasized that all scale dependent parameters, like the shoreline length, the shore development, the contour-line length and the lake bottom roughness have limited quantitative meaning if they are not normalized. The relationship between map scale, shoreline length ($1_o'$) and shore development (F') is illustrated in Table 13 with data from Lake Vänern.

F = 7.14 for Lake Vänern.

Table 13. Relationships between shore length $(1_0')$, shore development (F') and map scale (s) for Lake Vänern (from Håkanson, 1978a)

Scale factor (s)	Shore length $(1_0')$	Shore development (F')
10 000	2 007	7.38
50 000	1 875	6.89
60 000	1 893	6.96
180 000	1 642	6.03
250 000	1 608	5.91
500 000	1 325	4.87
1 000 000	1 012	3.72
Normalized values:	1 943	7.14

Lake bottom roughness (R, dimensionless); a measure of the degree of irregularity of the lake bottom. The R-value is only defined for whole lakes, accordingly:

$$R = \frac{0.165 \cdot (1_c + 2) \cdot \sum_{i=0}^{n} 1_i}{D_{50} \cdot \sqrt{a}}$$

(25)

where R = the normalized lake bottom roughness;
1_c = the contour-line interval in m;
1_i = the length of the given contour line in km;
D_{50} = the median depth in m;
a = the lake area in km^2.

Formula (25) may also be written as:

$$R = \frac{0.165 \cdot (1_c + 2) \cdot (1_0 + 1)}{D_{50} \cdot \sqrt{a}}$$

(26)

where l_o = the normalized shoreline length in km;

$$1 = \sum_{i=1}^{n} 1_i, \text{ i.e. } 1 = \text{ the total normalized length of the contour lines}$$

in km. The 1-value may be estimated by the L^+-value, equation (17).

To obtain comparable R-values for different lakes it is necessary to utilize a constant number of contour lines (n). The rule of thumb is that n should be equal to 5.

The lake bottom roughness (R) should not be used as a measure to compare the degree of bottom irregularity of various sub-basins in a lake. If this is required, then the form roughness (R_f) is a preferable measure.

The lake bottom roughness may also be determined directly from the echograms accordingly (Galvenius, 1975):

$$R' = \frac{1_x}{s_x} + \frac{1_y}{s_y} - 1 \tag{27}$$

where R' = the lake bottom roughness;

1_x, 1_y = the total length of the actual bottom profile along the x-direction and y-direction, respectively (in km);

s_x, s_y = the total straight-line length of the echosounded tracks in the x- and y-directions (in km).

No quantitative comparison between the R-value and the R'-value has yet been published.

The lake bottom roughness of Lake Vänern is comparatively large, R = 16. This is a high figure for a Swedish lake (see Table 14).

Form roughness (R_f, dimensionless); defined by the formula (Håkanson, 1974a):

$$R_f = \frac{0.165 \cdot (1_c + 2) \cdot \sum_{i=0}^{n} 1_i}{a} \tag{28}$$

The R_f-value may be utilized to quantitatively compare the degree of topographical irregularity of various sub-basins, e.g. in sediment sampling programs. Sub-areas with low form roughness may be sampled with less intensity than areas with great R_f-values, for example, when representative mean values of various sediment parameters are required.

Lake	Lake area (km^2)	R-value
Vänern	5 648	16.0
Vättern	1 856	7.1
Hjälmaren	484	12.0
Roxen	94.7	6.9
Ekoln	18.6	2.7
Saxen	7.1	7.3
Velen	2.8	8.3
Munksjön	1.1	4.7

Table 14. The normalized lake bottom roughness (R) for 8 Swedish lakes

Figure 32 illustrates the areal distribution of the form roughness in Lake Vänern. This map has been constructed in the following manner:

1. The working map has in this case been in scale 1:60 000 with 5 m contour-line interval.

2. The entire lake area has been divided into squares of 21.44 km^2.

3. The greatest and smallest water depth and the difference (ΔD) within each square has been determined.

4. Five contour lines have been drawn in each square. The contour-line interval (l_c) has been determined as the positive integer of the ratio $\Delta D/5$.

5. The total length of the contour lines within each square has been determined with CTP-technique.

6. The form roughness (R_f) for each square has been determined from equation (28).

7. Isopleths for the following R_f-values have been drawn by means of linear interpolation; R_f = 8, 16, 24 and 32.

Figure 32 illustrates in quantitative terms the areal distribution of the R_f-values in Lake Vänern. The map shows the relative differences in bottom irregularity within the lake area. The obtained R_f-value depends on, among other things, the choice of the size of the base area (see Fig. 33). Table 15 gives a summary of the areal distribution of the form roughness in Lake Vänern. From this table we can see that, for example, 13.7 % of the lake area has an

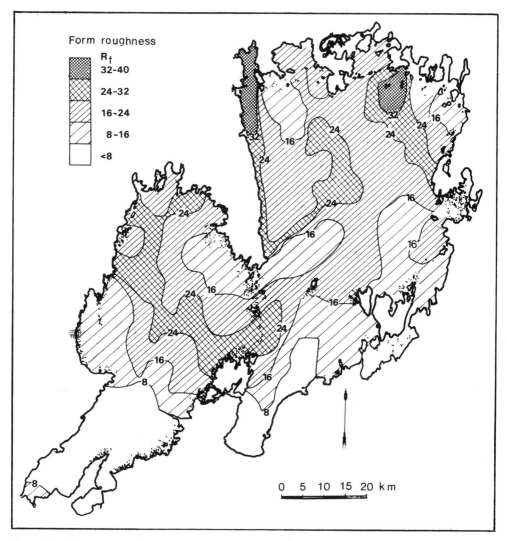

<u>Fig. 32.</u> The areal distribution of the form roughness in Lake Vänern

R_f-value <8. The median form roughness can be calculated to be 23.

It should be noted that the R_f-value is a relative measure which depends on, for example, the choice of the base area.

<u>Volume development</u> (V_d, dimensionless); is a measure used to illustrate the form of the lake basin. The V_d-value is defined as the quotient between the lake volume ($V = a \cdot \bar{D}$) and the volume of a cone whose base area is equal to the lake area (a) and whose height is equal to the maximum depth (D_{max}) of the lake, i.e.:

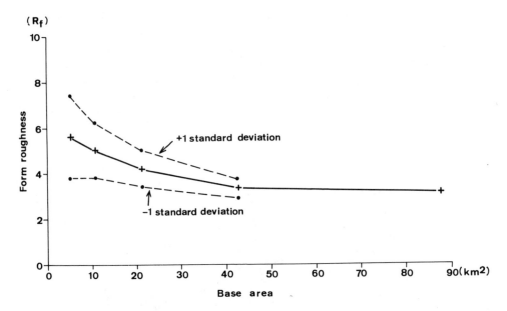

Fig. 33. The relationship between the form roughness (R_f) and the base area for which the R_f-value is determined. Example from a test area in Vänersborgsviken, Lake Vänern

Table 15. The areal distribution of the form roughness (R_f) in Lake Vänern

R_f-value	Area (km^2)	Area (%)
> 32	136	2.4
24 - 32	808	14.3
16 - 24	2 250	39.8
8 - 16	1 680	29.8
0 - 8	774	13.7

$$V_d = \frac{a \cdot \bar{D}}{1/3 \cdot D_{max} \cdot a} = \frac{3 \cdot \bar{D}}{D_{max}} \qquad (29)$$

V_d = 0.76 for Lake Vänern.

Islands, islets and rocks; the following classification may be used to define and distinguish islands, islets and rocks (Håkanson, 1974b):

56

Area (km²)	Name
<0.0001	rock
0.001 - 0.01	islet
0.01 - 1.0	island
>1.0	large island

This nomenclature has the advantage, before for example the one proposed by Hedenstierna (1948), that it is based on a specified area division. This simplifies statistical calculations. It also agrees, at least as well as any other classification system, with the prevalent linguistic usage.

The distribution of islands, islets and rocks in Lake Vänern is shown in Tables 16 and 17. The geographical positions of the given archipelago areas in Lake Vänern are illustrated on Fig. 34. The total number of identified objects on the Economic maps in scale 1:10 000 is 21 685, out of which 813 are islands and large islands. The total area of all objects is 244.86 km², out of which 232.37 km² (94.9 %) consists of islands and large islands.

Fresh-water archipelagos are quite rare on the earth and these areas in Lake Vänern make up a splendid recreation potential. 75 % of the tributary water passes through these archipelago areas before reaching the open water. In those rivers where the water contamination is high there is an evident conflict between recreation demands and water pollution.

Table 16. Islands, islets and rocks in the archipelago areas of Lake Vänern

Archipelago area	Rocks (<0.0001 km²)		Islets (0.0001-0.01 km²)		Islands (0.01-1.0 km²)		Large islands (>1 km²)	
	Number	Area	Number	Area	Number	Area	Number	Area
Vänersborgsviken	2 020	0.10	1 444	1.73	59	0.77	0	0
Eken-Kållandsö	1 660	0.08	1 187	1.45	95	4.14	2	9.29
Köpmannebro	1 200	0.06	855	1.07	56	2.51	0	0
Tössebäcken	510	0.02	366	0.60	53	4.39	0	0
Norra Dalbosjön	420	0.02	299	0.47	29	1.09	0	0
Millesvik	790	0.04	565	0.89	73	3.73	0	0
Lurö	840	0.04	600	0.31	75	3.67	1	1.84
Djurö	240	0.01	170	0.17	22	1.29	1	1.55
Mariestad	1 280	0.06	911	0.96	73	5.20	7	87.31
Segerstad	680	0.03	487	0.75	68	5.73	4	54.74
Karlstad	760	0.04	542	0.89	87	13.09	4	19.69
Kristinehamn	470	0.02	335	0.43	31	3.91	1	3.38
Åråsviken	1 110	0.05	794	1.32	53	3.05	0	0
Dalbosjön	6 470	0.31	4 756	6.27	372	17.28	2	9.29
Värmlandssjön	5 630	0.27	4 016	5.53	421	37.41	18	168.50
Vänern	12 100	0.58	8 772	11.80	793	54.69	20	177.79

Table 17. Rocks, islets and islands in Lake Vänern

Area (km^2)	Name	Dalbosjön		Värmlandssjön		Vänern	
		Number	Area (km^2)	Number	Area (km^2)	Number	Area (km^2)
< 0.0001	rocks	6 470	0.31	5 630	0.27	12 100	0.58
0.0001 - 0.01	islets	4 756	6.27	4 016	5.53	8 772	11.80
0.01 - 1	islands	372	17.28	421	37.42	793	54.69
> 1	large islands	2	9.29	18	168.50	20	177.79
Σ		11 600	33.14	10 085	211.72	21 685	244.86

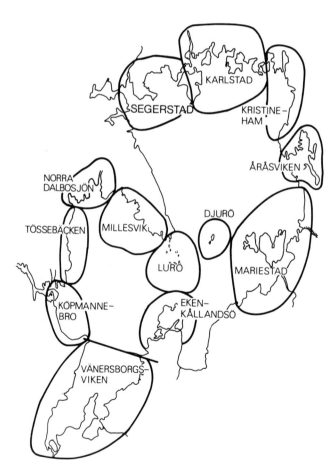

Fig. 34. Orientation map showing the different archipelago areas in Lake Vänern

Insulosity (I_n in %); defined as the percentage of the total lake area (A) that is occupied by islands, islets and rocks (A_i), i.e.:

$$I_n = \frac{A_i \cdot 100}{A} \qquad (30)$$

Insulosities over 30 % are very rare (Halbfass, 1922). The value is 4.2 % for Lake Vänern and 29 % for Lake Mälaren.

Profiles; are introduced to give a simple and illustrative picture of the bottom topography. The geographical position of the profile should always be given. A natural way to introduce profiles is by echograms, this may also give an indication about the character of the sediments, especially if low frequency echosounders are used. Low frequencies are, in this context, frequencies less than 30 kHz (see e.g. Müller, 1977).

Three echoprofiles from Lake Vänern are given in Figures 35, 36 and 37. The positions of these profiles are shown on Fig. 38. The echogram from Vänersborgsviken (Fig. 35) shows the shallow and undramatic bottom conditions that characterize this bay (and Kinneviken as well, compare Fig. 32). The sediments in this bay are dominated by granular materials (sand) with a comparatively low water content. The water contents of surficial sediments (0 - 1 cm, W_{0-1}) are given for some defined sample sites along the profile. The bottom areas in Vänersborgsviken are characterized by erosion and transportation processes (see Fig. 39). This implies that, e.g. dumping of dredging material could not be recommended in Vänersborgsviken. This was done in 1975 and practically none of the material remains on the site where it was dumped. The content of pollutants is very low in those erosion and transportation zones. Prac-

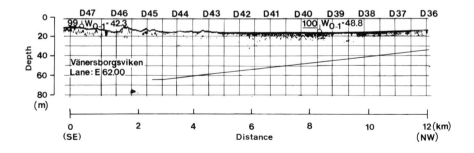

Fig. 35. Echoprofile from Vänersborgsviken, Lake Vänern. Decca lane: E 62.00; 99 = sample site (see Håkanson, 1977c), W_{0-1} = water content of surficial sediments (in % of the wet substance)

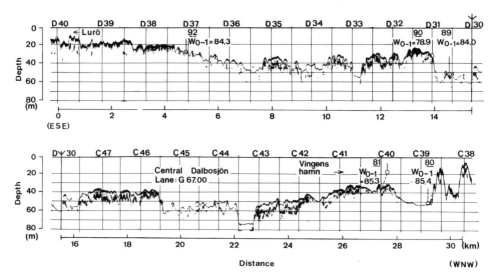

Fig. 36. Echoprofile from central Dalbosjön, Lake Vänern. Decca lane: G 67.00

tically all contamination is found within the accumulation areas, where fine material (medium silt and finer) continuously is deposited (see Table 18 and Fig. 40).

Table 18. The relationship between bottom dynamics (erosion-transportation-accumulation) and the physical and chemical status of the surficial sediments (0 - 1 cm) of Lake Vänern. Mean values and standard deviations (in brackets). From Håkanson (1977c)

			Erosion	Transportation	Accumulation
	Number of analyses		14	16	84
PHYSICAL STATUS	Water depth, m		14.9 (9.7)	22.6 (10.3)	42.8 (22.5
	Water content, % ws		40.3 (14.5)	64.0 (17.9)	84.0 (10.9)
	Bulk density, g/cm³		1.58 (0.24)	1.27 (0.19)	1.10 (0.08)
	Organic content, % ds		1.5 (0.8)	4.5 (2.3)	9.7 (5.3)
CHEMICAL STATUS	Nutrients, mg/g ds	N	0.78 (0.33)	1.14 (0.46)	2.7 (0.53)
		P	0.30 (0.24)	1.07 (0.80)	1.5 (1.0)
	Contaminating	Hg	0.026 (0.028)	0.27 (0.48)	1.34 (2.13)
	elements, ppm ds	Cd	<<1	0.7 (0.5)	1.4 (0.8)
		Cu	13 (9)	26 (10)	30 (7)
		Pb	34 (17)	64 (36)	104 (34)
		Zn	110 (63)	240 (125)	430 (210)
		Ni	22 (9)	31 (12)	28 (7)
	Conservative	Ag	<1	<1	<1
	elements, ppm ds	B	<20	<20	<20
		Be	<1	<1	<1
		Bi	<10	<10	<10
		Cr	<50	<50	⩽50
		Mo	<20	<20	<20
		Sn	<20	<20	<20
		V	110 (46)	150 (47)	150 (29)

60

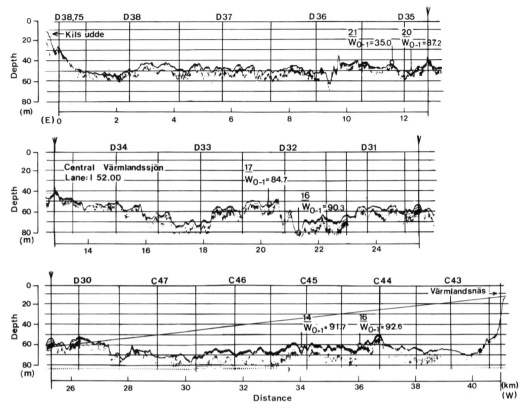

Fig. 37. Echoprofile from central Värmlandssjön, Lake Vänern. Decca lane: I 52.00

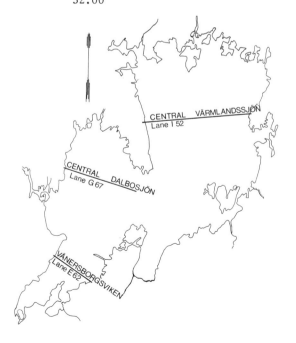

Fig. 38. The position of the given echoprofiles from Lake Vänern

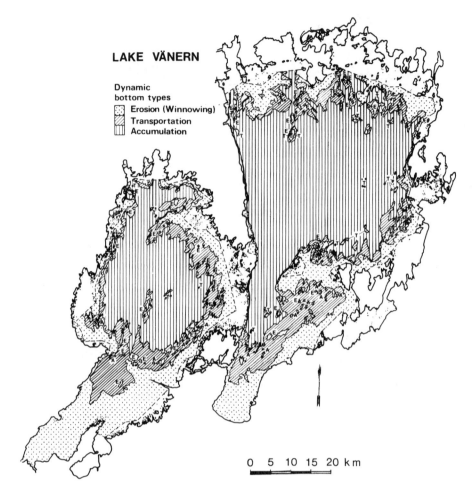

LAKE VÄNERN

Dynamic
bottom types
- Erosion (Winnowing)
- Transportation
- Accumulation

0 5 10 15 20 km

Fig. 39. The areal distribution of dynamic bottom types in Lake Vänern

The echogram from the central part of Dalbosjön (Fig. 36, Decca-lane G
67.00) illustrates the rough and irregular bottom topography that characteriz-
es Lake Vänern. The morphology is quite similar in the western main basin,
Värmlandssjön (Fig. 37, Decca-lane I 52.00). For further details about the
morphology of the Lake Vänern basin see Fredén (1978) and Lindh et al. (1978).

It should be noted that there is always a considerable height exaggeration
in echoprofiles, which implies that the bottom looks rougher than it really is.

The slope curve; gives the slope (α_p or α^o) on the vertical axis (positive ordinate) and the cumulative area on the horizontal axis (abscissa). The slope
curve can be established accordingly (from a working map in scale 1:60 000
with 5 m contour-line interval of Lake Vänern):

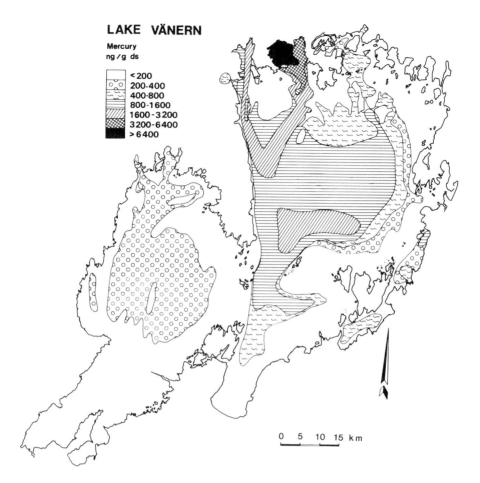

LAKE VÄNERN

Mercury
ng/g ds

	< 200
	200-400
	400-800
	800-1600
	1600-3200
	3200-6400
	> 6400

0   5   10 15 km

Fig. 40. The areal distribution of the main contaminant, mercury, in the surficial sediments (0 - 1 cm) of Lake Vänern. The dominating mercury
source is a chlor-alkali industry in northern Värmlandssjön, at Skog-
hall (from Håkanson, 1977c)

1. Slopes >10 %, i.e. α_p >10, occur in areas where the distance between the
contour lines is <0.83 mm. All areas on the map limited by such distances
have been marked and planimetrated. An alternative method to determine these
areas would be to use a quantimeter, e.g. Quantimet 720, from Image Analys-
ing Computers.

2. The total area limited by inclinations >10 % in Lake Vänern has been determined to be 84.72 km^2; i.e. 98.5 % of the lake area (100 - 84.72/5 648)
inclines <10 %.

3. The relationship between slope and area has then been determined in the
same manner for the following α_p-values: 8, 6, 4, 2, 1 and 0.5. The result
is illustrated in Fig. 31.

63

The median slope of the entire lake area (α_{50}) is determined from the slope curve. This value is 0.75 % in Lake Vänern. From the slope curve we may also, for example, see that only 10 % of the area inclines more than 5 % in Lake Vänern.

The hypsographic curve; (= the depth-area curve); is constructed by putting the depth on the negative ordinate and the cumulative area on the positive abscissa. The hypsographic curve represents certain elements of the form of the basin and it provides a means whereby the area of any depth level may be determined. The hypsographic curve may also be used in graphic determinations of the lake volume.

Hypsographic curves for Lake Vänern and its two main basins Dalbosjön and Värmlandssjön are illustrated on Fig. 41.

The percentage hypsographic curve; is obtained by putting the cumulative area in percentage on the positive abscissa and the depth on the negative ordinate. The percentage hypsographic curves for Lake Vänern, Dalbosjön and Värmlandssjön are given in Fig. 42.

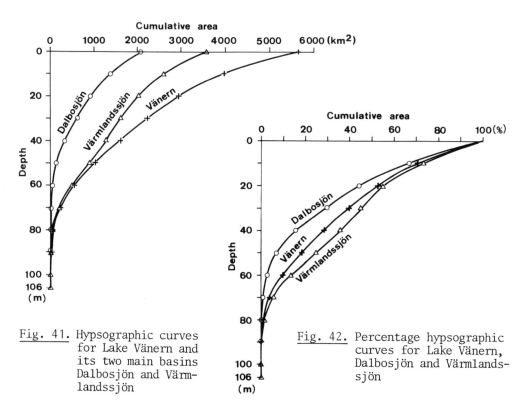

Fig. 41. Hypsographic curves for Lake Vänern and its two main basins Dalbosjön and Värmlandssjön

Fig. 42. Percentage hypsographic curves for Lake Vänern, Dalbosjön and Värmlandssjön

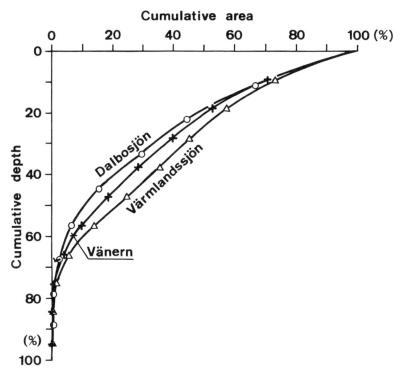

Fig. 43. Relative hypsographic curves for Lake Vänern, Dalbosjön and Värmlandssjön

The relative hypsographic curve; is obtained by putting depth as well as cumulative area in percentage on the two axes. The relative hypsographic curve
illustrates the lake form and nothing but the lake form. It is the base for
the definition of the lake form concept.

The relative hypsographic curves for Lake Vänern, Dalbosjön and Värmlandssjön are shown on Fig. 43.

The volume curve; illustrates the relationship between depth and volume in
the same way as the hypsographic curve shows the relationship between depth
and area. The cumulative volume at each depth level is given on the positive
abscissa and the depth on the negative ordinate.

The volume curves for Lake Vänern, Dalbosjön and Värmlandssjön are illustrated on Fig. 44.

The percentage volume curve; is obtained by putting the cumulative volume in
percentage on the positive abscissa and the depth on the negative ordinate
(see Fig. 45).

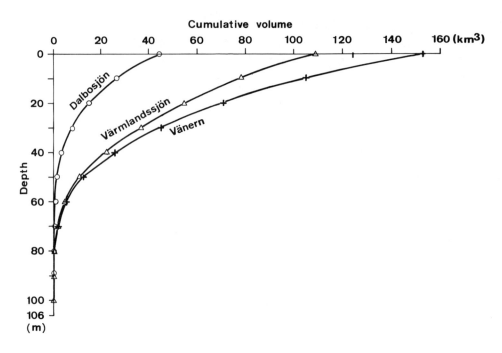

Fig. 44. Volume curves for Lake Vänern, Dalbosjön and Värmlandssjön

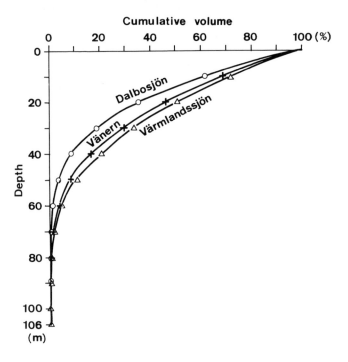

Fig. 45. Percentage volume curves for Lake Vänern, Dalbosjön and Värmlandssjön

The relative volume curve; is obtained by putting cumulative volume as well as cumulative depth in percentage on the two axes.

The relative volume curves for Lake Vänern and its two main basins are given on Fig. 46.

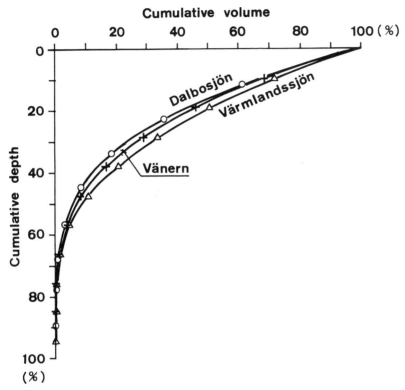

Fig. 46. Relative volume curves for Lake Vänern, Dalbosjön and Värmlandssjön

The lake form; is in this context defined and determined by the relative hypsographic curve (see Håkanson, 1977a). The mean lake form, labelled $f(\bar{x})$, signifies that there is a 50 % chance for an unknown lake to have a relative hypsographic curve above (on the convex side) or below (on the concave side) of the $f(\bar{x})$-curve. The statistical deviation forms corresponding to ±0.5, ±1.0, ±1.5, ±2.0 and ±3.0 standard deviations are labelled $f(\pm 0.5)$, $f(\pm 1.0)$ etc. Fig. 47 gives a schematical bathymetric interpretation of 4 different lake forms.

A lake with a relative hypsographic curve of the $f(-3.0)$-type has one (or more) areally limited deep hole(s), but is generally very shallow. A lake of

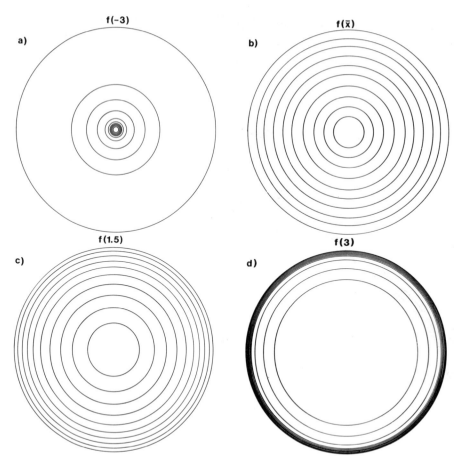

Fig. 47. Schematical bathymetric interpretation of four statistical lake
forms

the f(3.0)-type is trough-like with steep inclining walls and a very plane
areally dominating bottom. Terminology and class limits for the various lake
forms are illustrated in Fig. 48 and in Table 19. Very convex lakes, VCx, have
relative hypsographic curves between the limitation lines f(-3.0) and f(-1.5).
The probability that an arbitrary lake shall be of this VCx-type is only 6.5
%. Most lakes are of the convex (Cx) - 24.2 %, the slightly convex (SCx) -
38.3 %, or the linear (L) - 24.2 % type. Thus, the mean lake form, f($\bar{x}$), is
slightly convex (SCx). If there is no point of inflexion in the relative hypsographic curve, then this is denoted ma (= macro), e.g. Cxma, which is the
form of Lake Vänern. If the relative hypsographic curve has one (1) point of
inflexion, then this is designated by the label me(= meso), e.g. SCxme. If,
finally, there are two or more inflexion points, then the label mi (= micro)
is used, e.g. Lmi.

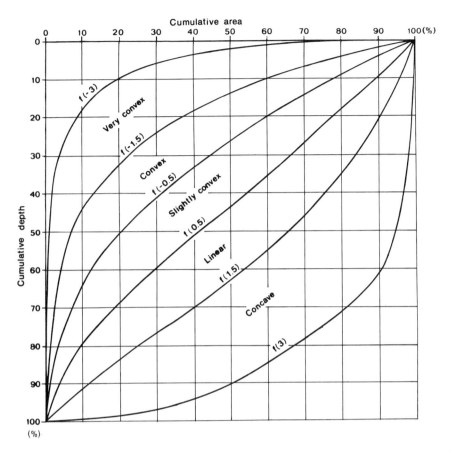

<u>Fig. 48.</u> Terminology and class limits for the classification system of lake forms

<u>Table 19.</u> Terminology, class limits and probability for various lake forms

Class limits	Name	Probability (%)
f(-3.0) - f(-1.5)	Very convex, VCx	6.5
f(-1.5) - f(-0.5)	Convex, Cx	24.2
f(-0.5) - f(0.5)	Slightly convex, SCx	38.3
f(0.5) - f(1.5)	Linear, L	24.2
f(1.5) - f(3.0)	Concave, C	6.5

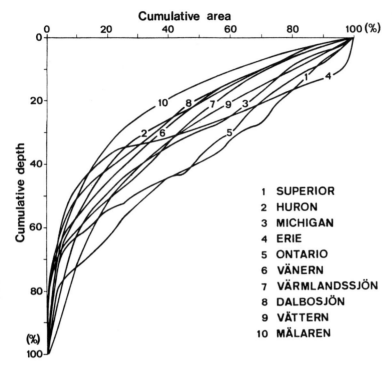

Fig. 49. Relative hypsographic curves for 10 large lakes

The relative hypsographic curves for 10 large lakes are shown on Fig. 49 and the forms of these lakes are given in Table 20. As already discussed, the lake form concept is of vital importance in the optimization model for lake hydrographic surveys and in lake volume determinations.

The lake type; in a geological perspective lakes may be looked upon as causal objects on the earth surface. As for other geomorphological features, one can say that lakes exist in three different stages - youth - maturity - old age. Lakes are generally created in connexion with some drastic geological event, e.g. vulcanism, tectonic activity or glaciation. The relief of the basin is under constant alteration by exogenic forces; weathering, erosion, transport and deposition. Practically all lakes in, for example, Scandinavia and Canada have been influenced by the enormous formcreating power of the last glacial era.

If no tectonic catastrophe happens to the Scandinavian should all Swedish lakes will, in due course, be filled up and vanish. Hutchinson (1957) has in a brilliant chapter discussed all major lake types, their origin, morphological character and distribution on the earth. He differentiates between

Lake	Form
Superior	SCxmi
Huron	Cxma
Michigan	SCxmi
Erie	Cx-SCxmi
Ontario	SCxmi
Vänern	Cxma
Värmlandssjön	Cxmi
Dalbosjön	Cxma
Vättern	SCxma
Mälaren	Cxma

Table 20. Lake forms of 10 studied lakes (from Håkanson, 1977a)

11 major lake types, which are divided into 76 sub-types. In this context, where the emphasis is on morphometry and not morphology, we will only brief-ly summarize the 11 major types:

1. Tectonic lakes, e.g. basins in graben between faults (Lake Bajkal and Lake Tanganyika).

2. Vulcanic lakes, e.g. maars, caldera lakes and lakes formed by damming of lava flows.

3. Landslide lakes, e.g. lakes held by rockslides, mudflows and screes.

4. Glacial lakes; a) lakes in direct contact with ice, e.g. lakes on or in ice and lakes dammed by ice;

    b) glacial rock basins, e.g. cirque lakes and fjord lakes;

    c) morainic and outwash lakes, e.g. lakes created by ter-minal, recessional or lateral moraines;

    d) drift basins, e.g. kettle lakes and thermokarst lakes.

5. Solution lakes; e.g. polje lakes and lakes formed in caves by solution and deposition of calcareous sinter.

6. Fluvial lakes; a) plunge-pool lakes;

    b) fluviatile dams, e.g. strath lakes, lateral lakes, del-ta lakes and meres;

    c) meander lakes, e.g. oxbow lakes and crescentic levee lakes.

7. Aeolian lakes, e.g. basins dammed by wind-blown sand and deflation basins.

8. Shoreline lakes, e.g. tombolo lakes and spits lakes.

9. Organic lakes, e.g. phytogenic dams and coral lakes.

10. Anthropogenic lakes, e.g. dams and excavations made by man.

11. Meteorite lakes, e.g. meteorite craters.

Lake Vänern is a glacial lake with several major fault zones (see Fredén, 1978 and Lindh et al., 1978). A summary of the morphometrical data for Lake Vänern is given in Table 21.

Table 21. Morphometrical data for Lake Vänern

Maximum length (L_{max}), km	141
Maximum effective length (L_e), km	97
Maximum width (B_{max}), km	72
Maximum effective width (B_e), km	57
Mean width ($\bar{B}$), km	40
Maximum depth (D_{max}), m	106
Mean depth ($\bar{D}$), m	27.0
Median depth (D_{50}), m	22.0
1:st quartile depth (D_{25}), m	43.5
3:rd quartile depth (D_{75}), m	8.5
Relative depth (D_r), %	0.12
Direction of major axis	SW - NE
Shoreline length (l_o), km	1 940
Total lake area (A), km^2	5 893
Lake area (a), km^2	5 648
Volume (V), km^3	153
Mean slope ($\bar{\alpha}_p$), %	2.3
Median slope (α_{50}), %	0.75
Shore development (F), dimensionless	7.14
Lake bottom roughness (R), dimensionless	16.0
Volume development (V_d), dimensionless	0.76
Island (>0.01 km^2), number	813
total area, km^2	232.49
Islets and rocks (<0.01 km^2), number	20 872
total area, km^2	12.38
Insulosity (I_n), %	4.2
Lake form	Cxma

5. ACKNOWLEDGEMENTS

I am greatly indebted to Nils Sjödin, SMHI, for valuable help and guidance concerning the sections on echosoundings and bathymetric map construction, and for providing the figures to these sections. I would also like to thank Anders and Ramona Källström for their constructive remarks and linguistic revising, Marianne Berglund for typing the manuscript and Kerstin Andersson for drawing the maps and the diagrams. The work has been sponsored by the National Swedish Environment Protection Board.

6. APPENDIX

Background data on Lake Vänern and its drainage area.

Geographical position	$58^O - 62^O$, 5 N
	11°, $5^O - 14^O$, 5 E
Drainage area	46 800 km^2
Within Norway, 18 %	7 400 km^2
Total lake area, 18.6 %	8 690 km^2
Total lake area, Lake Vänern excluded	3 040 km^2

Major lakes within the drainage area (a):

Femudden	201 km^2
Stora Le	137 km^2
Skagern	132 km^2
Glafsfjorden	102 km^2
Fryken lakes	102 km^2
Värmeln	79 km^2
Stora Lungen	61 km^2
Lelången	55 km^2

Deep lakes within the drainage area (D_{max}):

Femudden	131 m
Middle Fryken	120 m
Upper Fryken	110 m
Stora Le	102 m
Skagern	77 m
Rådasjön	76 m
Djupsjön	71 m
Stora Ullen	71 m

Major tributaries (water discharge):

Klarälven	166 m^3/sec.
Gullspångsälven	63 m^3/sec.
Byälven	61 m^3/sec.
Upperudsälven	45 m^3/sec.

Mean water discharge at the outlet:
   (Göta älv)   544 m^3/sec.

Retention time of the water in Lake Vänern	8.8 years
Mean surface run-off	11.6 l/sec., km^2
Land uprise at Karlstad	35 cm/100 years
Land uprise at Vargön	26 cm/100 years
Highest high-water level	+44.85 m (above sea-level)
Lowest low-water level	+43.16 m (above sea-level)
Difference between highest and lowest level from the mean water level	±0.85 m
Highest peak within the drainage area (Randalssølen in Norway)	+1 755 m
Mean yearly air temperature	approx. 5 $^{\circ}$C
Mean yearly precipitation	
on lake surface	approx. 550 mm
on land surface	approx. 700 mm
Precipitation distribution	summer max.
	spring min.
Dominating wind direction	SW
Evaporation	200-400 mm/year
Wood land	50 %
Rocks	granite, gneiss

Lake Vänern data:

Secchi disc transparency	4.0 m
pH	7.1
Total N	0.7 mg/l
Total P	0.008 mg/l
Conductivity, 20 $^{\circ}$C ms/m	8.0
Chlorophyll a	2 mg/m^3
Fish (most important catch)	vendace
	pike
	burbot
Catch	1.0 kg/ha

7. REFERENCES

Ahl, T., 1975. Effects of man-induced and natural loading of phosphorus and nitrogen on the large Swedish lakes. - Internationale Vereinigung für Theoretische und Angewandte Limnologie 19.

Beach Erosion Board, 1972. Waves in inland reservoirs. - Technical Memoir 132, Beach Erosion Corps of Engineers, Washington, D.C.

Eberly, W.R., 1964. Further studies on the metalimnetic oxygen maximum, with special reference to its occurrence throughout the world. - Investigations of Indiana Lake Streams, Volume 6.

Fredén, C., 1978. Vänerns kvartära utveckling. In: Håkanson, L., Vänerns morfometri och morfologi. - LiberFörlag, Stockholm.

Galvenius, G., 1975. Bestämning av sjöbottnens brutenhet ur ekogram. - Stencil, Tekniska Högskolan, Stockholm.

Gilbert, R., 1975. Sedimentation in Lillolet Lake, British Columbia. - Canadian Journal of Earth Sciences, Volume 12, No. 10.

Håkanson, L., 1974a. A mathematical model for establishing numerical values of topographical roughness for lake bottoms. - Geografiska Annaler. Volume 56 A, Häfte 3 - 4.

-       1974b. Inventering av öar i Vänern. - Statens Naturvårdsverk, SNV PM 529.

-       1977a. On lake form, lake volume and lake hypsographic survey. - Geografiska Annaler, Volume 59 A.

-       1977b. The influence of wind, fetch, and water depth on the distribution of sediments in Lake Vänern, Sweden. - Canadian Journal of Earth Sciences, Volume 14, No. 3.

-       1977c. Sediments as indicators of contamination; investigations in the four largest Swedish lakes. - SNV PM 839/Naturvårdsverkets limnologiska undersökning 92, Uppsala.

-       1978a. The length of closed geomorphic lines. - Mathematical Geology, Volume 10, No. 2.

-       1978b. Optimization of lake hydrographic surveys. - Water Resources Research. Volume 14, No. 4.

-       1978c. Vänerns morfometri och morfologi. - LiberFörlag, Stockholm.

Halbfass, W., 1922. Die Seen der Erde. - Petermanns Mitteilungen, Ergänzungsheft 185.

Hedenstierna, B., 1948. Stockholms skärgård. - Geografiska Annaler. Häfte 1 - 2.

Hutchinson, G.E., 1957. A treatise on limnology, Volume 1, Geography, Physics and Chemistry. - John Wiley and Sons, Inc., New York.

Lidholm, B., 1956. Venerns Seglationsstyrelse, en historik. - Affärstryckeriet i Lidköping AB.

Lindh, A., Ronge, B. and Stigh, J., 1978. Vänerns geologi. In: Håkanson, L., 1978a, Vänerns morfometri och morfologi. - LiberFörlag, Stockholm.

Mortimer, C.H., 1952. Water movements in lakes during summer stratification; evidence from the distribution of temperature in Windermere. - Philosophical Transactions of the Royal Society of London. Series B, Biological Sciences, No. 635, Volume 236.

Müller, H.E., 1977. Observations of interactions between water and sediment with a 30 kHz sediment echosounder. - Interaction between sediments and water. Proceedings of an international symposium held at Amsterdam, the Netherlands, September 6 - 10, 1976.

Norrman, J.O., 1964. Lake Vättern. Investigations on shore and bottom morphology. - Geografiska Annaler, Häfte 1 - 2.

Rawson, D.S., 1955. Morphometry as a dominant factor in the productority of large lakes. - Internationale Vereinigung für Theoretische und Angewandte Limnologie 12.

Sturm, M., 1975. Depositional and erosional sedimentary features in a turbidity current controlled basin (Lake Brienz). - IXth International Congress of Sedimentology Nice 1975, theme 5, Volume 5/2.

Vollenweider, R.A., 1968. Scientific fundamentals of the eutrophication of lakes and flowing waters, with particular reference to nitrogen and phosphorus as factors of eutrophication. - OECD, DAS/CSI, Paris.

Welch, P.S., 1948. Limnological methods. - The Blakiston Co., Toronto.

Wetzel, R.G., 1975. Limnology. - W.B. Saunders Co., London.

Lakes
Chemistry, Geology,
Physics

Editor: A. Lerman
With contributions by numerous experts

1978. 206 figures, 61 tables. XI, 363 pages
ISBN 3-540-90322-4

"This is an attractive and unusual volume. It contains 11 chapters dealing with many of the topics currently of interest to nonbiological limnologists (and one would hope to biologists as well). Each chapter contains from three to more than ten pages of fundamental material followed by a more detailed discussion of recent advances. This format works exceptionally well, and I would like to see it employed more often.

The book is also remarkable in that most of the 19 authors consider themselves to be not limnologists but geochemists, sedimentary geologists, physical oceanographers, aquatic chemists, or mineralogists. ...this book will contribute to a greater appreciation by biologists of the chemical and physical aspects of limnology.

The chapters contain a wealth of information that is simply not available in such concise form anywhere else. Aquatic chemistry is discussed in useful chapters on perturbations caused by humans (**W. Stumm** and **P. Baccini**) and chemical modeling of lakes (**D. M. Imboden** and **A. Lerman**). ...The chapter by **Eugster** and **Hardie** brings together in one place much of the elegant work that Eugster and his associates have done on highly saline lakes in Africa and North America.

Kelts and **Hsü** present a useful review of the numerous processes that control carbonate sedimentation in lakes. They have used their own data from Lake Zurich to illustrate most of the major points made in their discussion. This is particularly appropriate because Nipkow's classic work on paleolimnology and sedimentology was done on the varied sediments of that lake. Scanning electron micrographs of selected layers in sediment cores from the lake are used to give one an intimate view of the diatom frustules and minerals deposited through time.

Chemical perturbations by aquatic ecosystems resulting from human activities are examined by **Stumm** and **Baccini** in a notably synthetic chapter. ...**Jones** and **Bowser** treat the mineralogy and related sediment chemistry of lakes in a long, scholoary chapter. They have made a special effort to organize and summarize the current literature in the field (over 230 references are cited). The chapter was written to interest students in the subject and to provide a basic guide to the study and interpretation of the chemical and mineralogical characteristics of sediments as well as to provide a good critical discussion of the relevant literature. ...the authors have produced chapters that fit nicely into the conceptual framework of the book. The figures and tables are generally excellent. ...the book is first-rate." *Science*

Springer-Verlag
Berlin
Heidelberg
New York

Pond Littoral Ecosystems

Structure and Functioning

Methods and Results of Quantitative Ecosystems Research in the Czechoslovakian IBP Wetland Project

Editors: D. Dykyjová, K. Květ

1978. 183 figures, 100 tables. XIV, 464 pages
(Ecological Studies, Volume 28)
ISBN 3-540-08569-6

Contents: General Ecology and Inventarization of Biotic Communities. – Environmental Factors in Fishpond Littorals. – Primary Production and Production Processes in Littoral Plant Communities. – Structure and Functioning of Algal Communities in Fishponds. – Decomposition Processes in the Fishpond Littoral. – Structure and Role of Animal Populations in Pond Littorals. – Effects of Fishpond Management on the Littoral Communities. Exploitation of Reed. – Conservation of Plant Communities and Waterfowl in Wetlands of Czechoslovakia.

Springer-Verlag
Berlin
Heidelberg
New York

This book contains the results of the Czechoslovakian contribution to the wetland studies of the "International Biological Programme" (1965–1974). A team of ecologists investigated the fishpond littoral ecosystems in two biogeographical regions of Central Europe comparing environmental characteristics and functioning of the biotic components in the Trebon basin, a UNESCO/MAB biosphere reserve in Bohemia (Hercynian region), and in the Lednice fishponds, a State nature reserve in Moravia (Pannonian region).

More general information includes structure, productivity, and production processes of the vegetation and its macrophytes and algae, decomposition processes and mineral nutrient regime in fishpond littorals, the role of certain animal populations, and management and conservation of fishpond littoral vegetation and waterfowl.